CYCLURA

NATURAL HISTORY, HUSBANDRY, AND CONSERVATION OF WEST INDIAN ROCK IGUANAS

CYCLURA

Natural History, Husbandry, and Conservation of West Indian Rock Iguanas

JEFFREY M. LEMM AND ALLISON C. ALBERTS
San Diego Zoo Global, San Diego, CA, USA

ELSEVIER

AMSTERDAM • BOSTON • HEIDELBERG • LONDON
NEW YORK • OXFORD • PARIS • SAN DIEGO
SAN FRANCISCO • SINGAPORE • SYDNEY • TOKYO

Academic Press is an imprint of Elsevier

AP

Academic Press is an imprint of Elsevier
32 Jamestown Road, London NW1 7BY, UK
225 Wyman Street, Waltham, MA 02451, USA
525 B Street, Suite 1800, San Diego, CA 92101-4495, USA

First edition 2012

Cover image
An adult male Turks and Caicos iguana (*Cyclura carinata carinata*) basks near a wrecked Haitian boat on French Cay in the Turks and Caicos Islands.

British Library Cataloguing-in-Publication Data
A catalogue record for this book is available from the British Library

Library of Congress Cataloging-in-Publication Data
A catalog record for this book is available from the Library of Congress

ISBN: 978-1-43-773516-1

For information on all Academic Press publications
visit our website at www.elsevierdirect.com

Typeset by TNQ Books and Journals

Printed and bound by CPI Group (UK) Ltd, Croydon, CR0 4YY

Contents

List of Contributors vii
Preface xi
Acknowledgments xv
About the Authors xvii

1. Evolution and Biogeography

CATHERINE L. STEPHEN

Evolution on Islands 3
Cyclura's Wild Ride 5
What's in a Name? 9
A Consequence of Island Evolution 10

2. Species Accounts

Species of the Genus *Cyclura* 16
Turks and Caicos Iguana (*Cyclura carinata carinata*) 17
Jamaican Iguana (*Cyclura collei*) 21
Rhinoceros Iguana (*Cyclura cornuta cornuta*) 24
Mona Island Iguana (*Cyclura cornuta stejnegeri*) 28
Andros Island Iguana (*Cyclura cychlura cychlura*) 30
Exuma Island Iguana (*Cyclura cychlura figginsi*) 34
Allen Cays Iguana (*Cyclura cychlura inornata*) 38
Grand Cayman Blue Iguana (*Cyclura lewisi*) 42
Sister Isles Iguana (*Cyclura nubila caymanensis*) 46
Cuban Iguana (*Cyclura nubila nubila*) 49
Anegada Island Iguana (*Cyclura pinguis*) 54
Ricord's Iguana (*Cyclura ricordii*) 59
White Cay Iguana (*Cyclura rileyi cristata*) 62
Acklin's Iguana (*Cyclura rileyi nuchalis*) 65
San Salvador Iguana (*Cyclura rileyi rileyi*) 68
Navassa Island Iguana (*Cyclura cornuta onchiopsis*) – Extinct 71
Additional Notes From the Fossil Record 72
Possible Unnamed Species 72
Unnamed Species 72
Unknown Species 73

3. Natural History

Habitat Requirements and Home Range 77
Diet and Foraging 80

Predators and Defense 84
Social Behavior 87
Reproduction and Life History 89

4. Husbandry

Population Management 97
Quarantine 98
Housing 98
Feeding 104
Capture, Restraint, and Handling 109
Reproduction and Nesting 112
Hatchling Care 117
Record Keeping 123

5. Nutrition

ANN M. WARD, JANET L. DEMPSEY

Introduction 129
Feeding Ecology and Digestive Morphology 129
Nutrient Content of the Natural Diet 132
Nutrient Requirements 132
Food Availability and Practical Captive Diets 136
Seasonal Changes 139
Nutrition-Related Health Concerns 140
Vitamin D Needs and Assessing Vitamin D Status 141
Serum Vitamins 142

6. Health and Medical Management

NANCY P. LUNG

Introduction 147
Health and Medical Management of Captive Iguanas 147
Reproductive Problems 153
Clinical Techniques 161
Health of Free-Ranging and Headstart Rock Iguanas 166
Pathology 171

7. Conservation

Introduction 175
Threats 178
Conservation Actions 185
Outreach and Education 196
Long-Term Species Recovery Planning 199

Bibliography 201
Glossary 213
Index 217

List of Contributors

Janet L. Dempsey MS Senior Nutritionist, Technical Services, Nestle Purina PetCare Company, Saint Louis, MO

Nancy P. Lung DVM Director of Veterinary Services, Fort Worth Zoo, Fort Worth, TX

Catherine L. Stephen PhD Assistant Professor, Utah Valley State College, Alpine, UT

Ann M. Ward MS Director of Nutritional Services, Fort Worth Zoo, Fort Worth, TX

We dedicate this book to our families: Jeff to wife Carolyn, sons Matthias and Garrett, and daughter Sydney, and Allison to husband Michael and sons Connor and Jonathan, for their love, support, and daily inspiration in all we do.

Preface

Ricord's iguana (*Cyclura ricordii*).

The West Indian rock iguanas of the genus *Cyclura* are a unique group of large lizards that inhabit many of the islands of the Bahamas, Greater Antilles, and Lesser Antilles. Until recent times, these iguanas were the largest of the native land animals on these islands. Because they are important seed dispersers, they help keep tropical dry forests and other biomes in the region healthy (Iverson, 1985; Alberts, 2000).

As a group, rock iguanas are considered to be the most endangered lizards in the world, primarily because much of their fragile island habitat has been destroyed by human encroachment. In addition, non-native livestock have severely degraded the land and food plants on which the iguanas rely for survival. Feral predators such as cats, mongooses, and dogs prey on rock iguanas, especially juveniles, and many taxa of rock iguanas are in danger of becoming extinct. The loss of rock iguanas has serious consequences for the ecosystems in which they live (Alberts, 2000).

Researchers and conservationists have been working diligently to save rock iguanas for the past several decades. Information on their natural history and the threats facing each taxon is being used to develop and implement comprehensive management plans for the most endangered species. Removal of feral and exotic species, headstarting and captive breeding efforts, translocations, establishment of preserve systems, and education and capacity building programs for local people have served to help restore and protect many rock iguana species and the environments in which they live.

This book represents an attempt to summarize the wealth of knowledge found in scientific papers and other writings by academic scientists, professional conservation managers, and non-academic laypeople concerned about the plight of rock iguanas. This is the first publication to discuss natural history and captive maintenance of these lizards in combination with scientific research and conservation.

Taxonomy and genetics are scientific tools that help increase knowledge of species and their relationships, but they can also aid in wildlife conservation. There have been many changes in the taxonomy of rock iguanas over the past two decades, and there will undoubtedly be more to come. The chapter contributed by leading expert Dr. Catherine Stephen on rock iguana taxonomy outlines the evolutionary relationships among these marvelous lizards. The species accounts were written in an attempt to provide as much natural history information as possible for each taxon, and to help interpret many of the scientific papers that have been written on these animals. The natural history chapter provides a general overview of how rock iguanas live and the adaptations they have acquired to survive in the harsh landscapes in which they are found.

Grand Cayman Blue iguana (*Cyclura lewisi*) hatchling at Blue Iguana Recovery Program headstart and breeding facility on Grand Cayman.

Zoos and conservation organizations have strict breeding protocols for their rock iguanas. Genetic data are closely tracked and pairings are made to keep bloodlines as pure and unrelated as possible. Later, these captive-bred animals can be used to repopulate or augment populations in the wild that may need assistance. In order to breed rock iguanas, advanced captive husbandry skills are necessary, especially for species that are difficult to breed. The husbandry chapter in this book is a compilation of 20 years of captive maintenance and breeding work that has been heavily influenced by observing iguanas in the wild. It is hoped that the inclusion of captive maintenance guidelines will assist zoos that are eager to join the AZA (Association of Zoos and Aquariums) Rock Iguana Species Survival Program (SSP). Together with the IUCN (International Union for Conservation of Nature) Iguana Specialist Group, the Rock Iguana SSP manages captive rock iguanas in North American zoos, including breeding and conservation programs. In addition, the husbandry information contained in this book should be of value in assisting range country headstarting and captive breeding programs, as well as private individuals who seek proper protocols to care for their animals. Rock iguanas have become increasingly popular in the private sector over the years.

Nutrition is very important to the health of captive iguanas. In this book nutritionists Ann Ward and Janet Dempsey share their expertise on the diets of captive and wild rock iguanas. Another chapter, on health and medical management, by veterinarian Dr. Nancy Lung, discusses a wide variety of disease issues facing rock iguanas, both in captivity and in the wild.

Our goal in writing this book is to educate the public on the plight of rock iguanas and the unfortunate demise their populations have suffered at the hands of people and their feral animals. Fortunately, caring people from the zoos, conservation organizations, universities, the private sector, and other groups have come together to help conserve rock iguanas, and the final chapter of this book focuses on the myriad of conservation problems and the innovative solutions currently being developed to address them.

Cuban iguana (*Cyclura nubila nubila*) basking near oceanside cliffs at Guantanamo Bay, Cuba.

Acknowledgments

We thank our fellow members of the IUCN Iguana Specialist Group, without whose ongoing dedication and passion for rock iguanas this book would never have been possible. The International Iguana Foundation has been an ongoing source of financial and logistical support for rock iguana conservation and was a key supporter for much of the work described in this book.

All photos/drawings in this book are by Jeff Lemm unless otherwise noted. We thank all those who have kindly supplied illustrations.

About the Authors

Jeff Lemm is a herpetologist in the Applied Animal Ecology Division of the San Diego Zoo Institute for Conservation Research. Jeff has had a lifelong interest in reptiles and amphibians and has been working with West Indian rock iguanas for the past 20 years. He has traveled extensively and has photographed and/or studied almost every extant species of rock iguana in the wild. Jeff currently manages the husbandry and breeding program for West Indian iguanas at the San Diego Zoo and has been a member of the IUCN Iguana Specialist Group since its inception. When time (and family) permit, Jeff still enjoys working on conservation and research projects with wild iguanas.

Jeff Lemm and Gitmo, a large male Cuban iguana. *Photo © Bruce Farnsworth.*

Allison Alberts, PhD, currently serves San Diego Zoo Global as Chief Conservation and Research Officer. She is responsible for the ongoing conservation science activities at the San Diego Zoo and the San Diego Zoo Safari Park, including work at the Institute for Conservation Research and at field sites in 35 countries around the world. As a reptile and amphibian specialist, she has participated in conservation programs for endangered iguanas in Costa Rica, Cuba, the Turks and Caicos Islands, and Fiji, as well as working with komodo dragons, sea turtles, desert tortoises, and a variety of native California frogs, lizards, and snakes. Much of her research has focused on the development of innovative techniques for restoring critically endangered species to the wild. Her work has been recognized by the US Fish and Wildlife Service, the National Science Foundation, the American Association of Museums, and the Association of Zoos and Aquariums. She is also co-founder and past co-chair of the IUCN Iguana Specialist Group, and is currently president of the International Iguana Foundation.

Dr. Allison Alberts with a Cuban iguana named Sunny. *Photo by Ken Bohn/San Diego Zoo Global.*

CHAPTER

1

Evolution and Biogeography

CHAPTER

1

Evolution and Biogeography

Catherine L. Stephen

OUTLINE

Evolution on Islands	3	What's in a Name?	9
Cyclura's Wild Ride	5	A Consequence of Island Evolution	10

The geographic distribution of rock iguanas (*Cyclura*, Harlan, 1824) across the islands of the Caribbean shows an interesting pattern. Almost every species is relegated to a single island or island group. How did each species get to where it is today? Why are the species so morphologically distinct from one another? How old is each species? These and other questions can be addressed by combining available data on species relationships, distributions, geologic history of the Caribbean, and our knowledge of evolutionary processes.

It is helpful to understand a few of the fundamentals of island evolution when thinking about the evolutionary history of rock iguanas.

EVOLUTION ON ISLANDS

Biologists have spent much time studying the processes that shape the variation between island taxa. There are famous examples of fantastic species diversity along island archipelagos: the Galapagos has the mockingbirds, finches, and tortoises; in Hawaii are the honeycreepers, fruit flies, and silverswords; the Greater Antilles in the Caribbean host the *Anolis* radiation, as well as curly tail lizards, hutias, and rock iguanas.

Often the first step in the process of diversification is isolation of individuals from their parent population. This can happen in myriad ways, two of which are quite common for

Cyclura, *First Edition* DOI: 10.1016/B978-1-4377-3516-1.10001-9

island species: vicariance and dispersal. Vicariance occurs when a barrier arises that physically isolates a population from others of its species, such as tectonic plate separation. Diversification via dispersal begins when several organisms of a species share a rare event (e.g., rafting on a mat of vegetation to a new, distant location) leading to the establishment of a new population (Figure 1.1).

By either means, the newly isolated population is now subject to a novel set of natural selection pressures, such as differences in the climate, animal and plant communities, food resources, substrate, etc. As land masses break up, what was once an inland prairie can become a coastal area, and be subject to a vastly different microclimate regime. When sea level rises, lowland areas become submerged and populations unable to disperse across water are isolated on higher ground. Changes in ocean circulation will change temperature and humidity, soil composition, and vegetation. Some members of the biological community will be able to thrive in the new conditions, while others will dwindle. Subsequent changes in community structure will affect species—species interactions and resource availability. In every generation, genetic mutations will arise in these isolated populations that can affect such traits as behavior, morphology, physiology, and communication. Such mutations might

FIGURE 1.1 Grand Cayman Blue iguanas are believed to be relatives of iguanas that rafted to the Cayman Islands from Cuba.

facilitate a better fit to the changed structural environment and community composition, and these become the drivers of evolution. Over many thousands of generations, without continued genetic connection to the ancestral population, increased differentiation will eventually lead to diversification and speciation.

CYCLURA'S WILD RIDE

The Greater Antilles island group originated on the edge of the Caribbean Plate in the Pacific Ocean and pushed its way eastward to its current position in the Atlantic Ocean. Current scientific data support the scenario that much of the Greater Antilles' flora and fauna, potentially including the ancestor to *Cyclura*, colonized the islands when a connection with South America existed via the Aves Ridge during the Eocene–Oligocene transition 35–33 million years ago (mya) (Malone et al., 2000; Hedges, 2001; MacPhee et al., 2003; Iturralde-Vinent, 2006). Subsequently, the biological communities on these lands experienced land mass connections and disconnections, island emergences and submergences, sea level fluctuations, and literally millions of tropical storms. The combination of geologic, genetic, and evolutionary data allows us to reconstruct a likely sequence of events that led to the modern day diversity of the rock iguanas (Figures 1.2 and 1.3). However, the evolutionary scenario presented here is just one of several hypotheses, and new information could alter our understanding of the evolution of *Cyclura* iguanas.

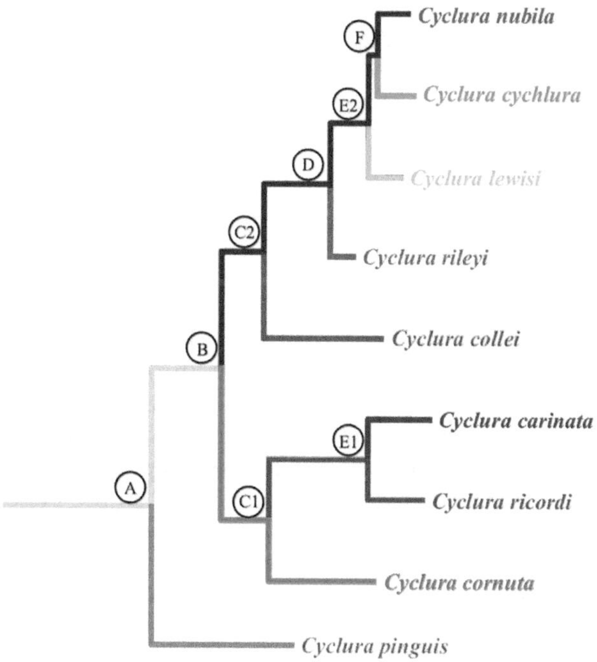

FIGURE 1.2 Evolutionary relationships of the *Cyclura* species. Letters mark each divergence (node). Each species is color coded to match their distribution in Figure 1.3.

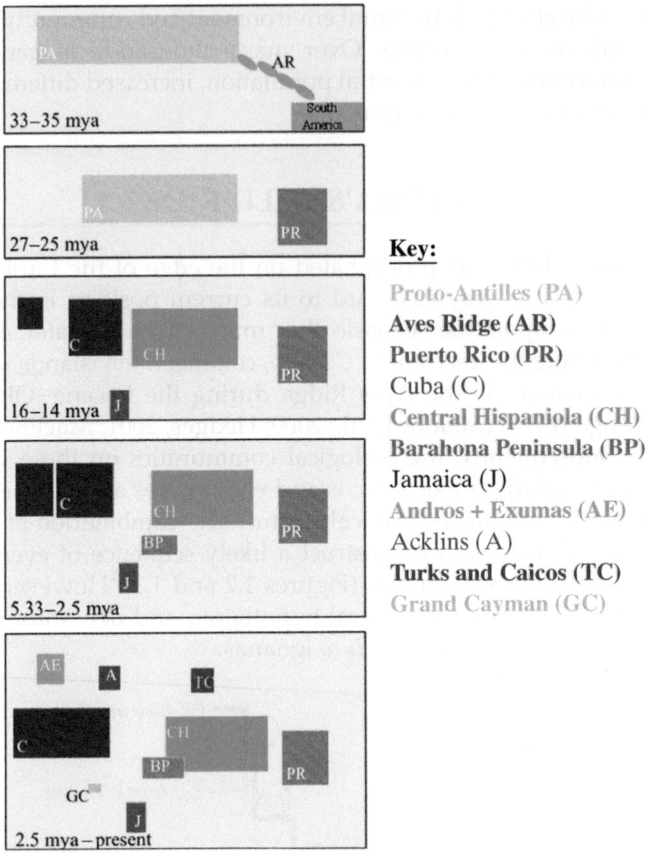

FIGURE 1.3 A chronological series of some of the geologic changes in the Caribbean from the Oligocene–Eocene transition to the present (based on Iturralde-Vinent, 2006). Each island (or island group) is color coded to indicate the *Cyclura* species occurring there.

In the earliest configuration of the "proto-Antillean" land mass, the three largest islands in this group were connected (Lewis and Draper, 1990; Iturralde-Vinent, 2006; Pindell et al., 2006), and the iguanid colonizer was apparently successful enough to disperse across this land mass. The initial divergences within *Cyclura* resulted when this land mass broke up due to tectonic activity and deep waterways formed, separating groups of animals. The earliest new species line to be isolated became *C. pinguis* on the Puerto Rican Bank (node A). This species, the most basal (oldest) of the genus, was once present throughout the Puerto Rican island bank, but is now restricted to a few small islands in the British Virgin Islands. The remaining island mass then separated into two parts (node B), eastern Cuba and central Hispaniola (excluding the southern peninsula), with the formation of the Windward Passage waterway between the two islands. Cuba's successful *Cyclura* population gave rise to many descendent species as they dispersed onto other islands (see beyond), while the core Cuban population evolved into the species *C. nubila*. Originally, Hispaniola (today comprised of the

FIGURE 1.4 Hispaniola is the only island in the world to contain sympatric rock iguana species; the Ricord's iguana (seen here) and the Rhinoceros iguana.

countries of Dominican Republic and Haiti) had only the population whose modern descendents are *C. cornuta.*

Hispaniola has the distinction of being the only island to contain two species of rock iguanas today (Figure 1.4). *C. cornuta* is distributed throughout most of the island, whereas *C. ricordii* is restricted to the Barahona Peninsula in the arid Neiba Valley and coastal lowlands. The two species are sympatric (utilize the same habitat) throughout the limited range of *C. ricordii.* So, how did they diverge in the first place? Somewhere between 2 million years ago and 100,000 years ago, the Baharona Peninsula was joined with the rest of Hispaniola, forming the Neiba Valley. One hypothesis is that the ancestors of *C. ricordii* colonized this island-peninsula (node C1) at some point after it was no longer submerged and evolved differentiating traits prior to the island's connection with Hispaniola (thereby forming the peninsula). This lineage also gave rise to the species *C. carinata* (node E1), found throughout the Turks and Caicos Islands, the southernmost islands in the Bahamas chain (Figure 1.5). This likely resulted from a raft of individuals being swept off of the island-peninsula and through the Windward Passage.

The subsequent *Cyclura* lineages originated through overwater rafting from Cuban ancestors. According to the genetic data, at nearly the same time the most common ancestors to *C. ricordii* and *C. cornuta* diverged, Jamaica was colonized (node C2). This event was followed by establishment of the *C. rileyi* lineage in the southern Bahamas (node D). Genetic data further suggest that the current distribution of *C. rileyi* on the separate island banks of Acklins, San Salvador, and Exuma is the result of relatively recent dispersal (Malone et al.,

FIGURE 1.5 The Turks and Caicos iguana (seen here) is believed to have evolved from the same lineage as the Ricord's iguana.

FIGURE 1.6 The Andros iguana is one of many species of rock iguanas derived from Cuban ancestors.

2000). However, there has not been a fine scale genetic study of *C. rileyi*, which is needed to pinpoint the original population. Given its proximity to Cuba and the prevailing hurricane paths, it is suspected that the population of origin may have been Acklins Island. Next, settling of the ancestors to *C. lewisi* on Grand Cayman (node E2) occurred at about the same time iguanas landed on the Turks and Caicos Islands. The final species to differentiate from the Cuban core population was *C. cychlura*, when rock iguanas arrived in the northern Bahamas (node F) (Figure 1.6).

The Pleistocene glaciation cycle had a lasting effect on the diversity of *C. cychlura* species in the Bahamas. During the height of glaciation, when the water level was much lower (Haq et al., 1987), Andros Island and the Exuma Cays were connected by dry land. Analysis of genetic data (Malone et al., 2003) indicates that during this time *C. cychlura* had many interconnected populations throughout this region. When water level rose and the islands became isolated, populations on the Exumas became separated from those on Andros Island and, subsequently, from one another. The result is a species made up of one large population and many very small, isolated populations that are genetically distinctive (Malone et al., 2003).

WHAT'S IN A NAME?

Species definitions come in very many shapes and sizes and are often dependent upon the data being used in the designation. Fortunately, there are some conceptual commonalities across the more frequently used definitions:

1. Each species is on an evolutionary trajectory separate from other species.
2. Populations of a species are bound through time via the processes of gene flow and heredity.
3. A species has accumulated measurable changes (evolved) that both unify its populations and make it distinct from the ancestral species.The nomenclature will ideally reflect the evolutionary history of the species.

The taxonomy of *Cyclura* has been relatively stable compared to many other reptile groups (e.g., *Coluber/Masticophis, Cnemidophorus/Aspidoscelus*). The reason for this is twofold:

1. Once a *Cyclura* population became isolated from its ancestral population, there was little to no "leakage" in the gene pool through migrations between the new and original population, allowing for clear morphological and genetic diversification over time.
2. The geographic boundaries of each species are easily demarcated by the island or island bank that it occupies.

However, there have been two recent nomenclatural changes, both of which resulted from scientific understanding of new genetic data. In 2004, *C. lewisi* on Grand Cayman was elevated to full species status (Burton, 2004a), based mainly on its highly unique DNA lineage and its placement on the *Cyclura* family tree (Malone et al., 2000). According to mitochondrial DNA data, the Grand Cayman population was established (likely from a Cuban lineage) earlier than the *C. cychlura* populations were established on the Bahaman bank (also from a Cuban lineage). Since time to common ancestry determines the degree to which two groups are related, *C. nubila* is more closely related to *C. cychlura* than it is to the

FIGURE 1.7 The Cuban iguana was at one time thought to be most closely related to the iguanas of the Cayman Islands. Recent genetic work has found that Cuban iguanas are actually more closely related to Bahamian iguanas of the species *cychlura*.

population on Grand Cayman (Figure 1.7). Data currently being collected from nuclear DNA of *C. lewisi* so far support its elevation to species status (C. Stephen and L. Buckley, unpublished data).

The most recent nomenclature change involved the Booby Cay population of rock iguanas. The population was originally designated as a separate subspecies, *C. carinata bartschi,* because its location on the Mayaguana bank in the Bahamas is very distant from its conspecifics in the Turks and Caicos Islands. However, using both mitochondrial and nuclear DNA data, the population was determined to be very recently established (Bryan et al., 2007). In fact, the population on Booby Cay shared a more recent common ancestor with populations on the Caicos bank than the Caicos populations did with those on the Turks bank. Anecdotal evidence suggests that iguanas were translocated to Booby Cay by humans, possibly as a reliable food source along the sugar cane trade route.

A CONSEQUENCE OF ISLAND EVOLUTION

One feature that most islands share is that they lack mammalian predators common on the mainland. Because most terrestrial mammals do not fare well on long distance oceanic

FIGURE 1.8 Feral dogs are common on many Caribbean Islands and often prey on iguanas.

crossings and require a large area to maintain a viable population, they seldom succeed in colonizing islands. This has dire consequences for endemic island species in that they can no longer recognize such predators as a threat. Charles Darwin's (admittedly very informal) experiment with marine iguanas drives home this point. Because they did not flee when approached, Darwin was easily able to pick them up and toss them into the ocean. The iguanas quickly scrambled back onto land into the waiting arms of their harasser, only to be tossed out again. While seemingly counterintuitive, this behavior is actually predictable when considered through the lens of evolution: the only predators on adult marine iguanas for millions of years existed in the waters of the ocean and, hence, they instinctively escaped from the water when they sensed danger.

Likewise, in their tens of millions of years of evolution, *Cyclura* have lost the mechanisms possessed by their mainland ancestors for responding to the threats posed by mammalian predators such as dogs, cats, mongooses, and humans (Figure 1.8). As a result, introduced mammals have the capacity to wreak havoc on *Cyclura* populations very quickly, and unfortunately have already decimated populations and even entire species. Control, or where possible elimination, of introduced mammalian predators is critical for the survival of remaining populations. Ongoing conservation efforts are essential to ensuring that today's *Cyclura*, the brilliant result of millions of years of evolution, can continue on their amazing evolutionary journey.

Acknowledgments

I am very grateful to Allison Alberts and Jeff Lemm for inviting me to contribute this chapter; it is a pleasure to work with such excellent people. Thanks to Allison Alberts, Jeff Lemm, Daniel Stephen, and Bruce Weissgold, who were all kind enough to lend their skills in editing the text and/or figures.

CHAPTER

2

Species Accounts

CHAPTER

2

Species Accounts

OUTLINE

Species of the Genus *Cyclura* 16

Turks and Caicos Iguana (*Cyclura carinata carinata*) 17

Jamaican Iguana (*Cyclura collei*) 21

Rhinoceros Iguana (*Cyclura cornuta cornuta*) 24

Mona Island Iguana (*Cyclura cornuta stejnegeri*) 28

Andros Island Iguana (*Cyclura cychlura cychlura*) 30

Exuma Island Iguana (*Cyclura cychlura figginsi*) 34

Allen Cays Iguana (*Cyclura cychlura inornata*) 38

Grand Cayman Blue Iguana (*Cyclura lewisi*) 42

Sister Isles Iguana (*Cyclura nubila caymanensis*) 46

Cuban Iguana (*Cyclura nubila nubila*) 49

Anegada Island Iguana (*Cyclura pinguis*) 54

Ricord's Iguana (*Cyclura ricordii*) 59

White Cay Iguana (*Cyclura rileyi cristata*) 62

Acklin's Iguana (*Cyclura rileyi nuchalis*) 65

San Salvador Iguana (*Cyclura rileyi rileyi*) 68

Navassa Island Iguana (*Cyclura cornuta onchiopsis*) – EXTINCT 71

Additional Notes From the Fossil Record 72

Possible Unnamed Species 72

Unnamed Species 72

Unknown Species 73

Cyclura, First Edition DOI: 10.1016/B978-1-4377-3516-1.10002-0

SPECIES OF THE GENUS *CYCLURA*

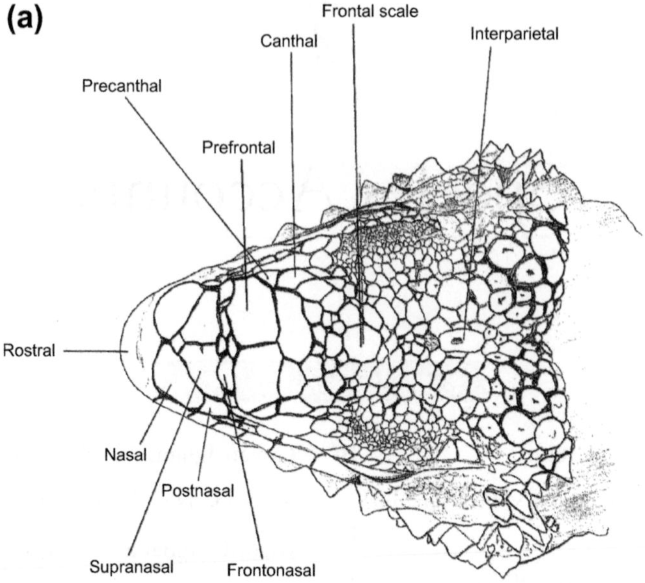

FIGURE 2.1(A) Head scalation (top) of a Cuban iguana (*Cyclura nubila nubila*).

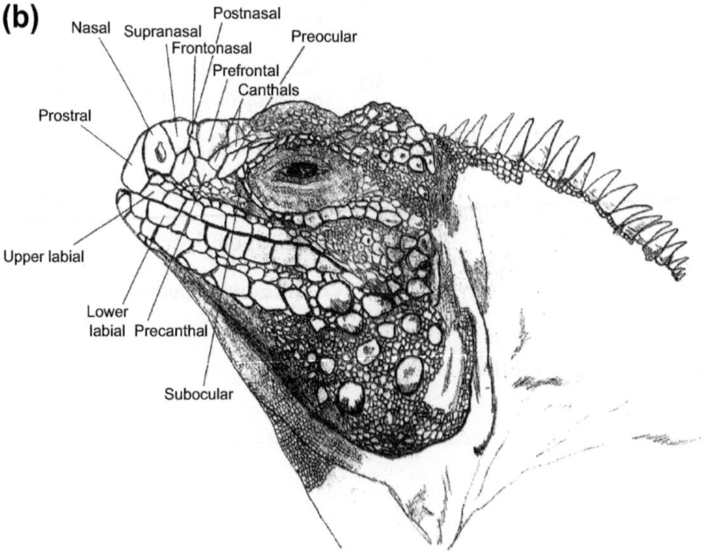

FIGURE 2.1(B) Head scalation (side) of a Cuban iguana (*C. n. nubila*).

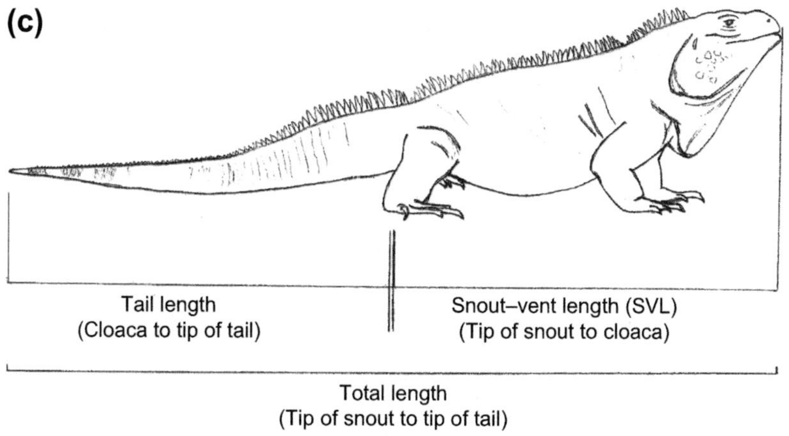

FIGURE 2.1(C) Body measurements of a rock iguana.

TURKS AND CAICOS IGUANA (*Cyclura carinata carinata*)

Harlan, 1824 Synonyms: *Iguana carinata*, Gray, 1831; *Cyclura carinata*, Harlan, 1824; *Cyclura carinata bartschi*, Cochran, 1931

Description

The Turks and Caicos iguana (*C. c. carinata*) (Figure 2.2) is a relatively small rock iguana that is characterized by a lack of enlarged scales on the upper surface of the head. It has 7–13 scales between the anterior canthal scales, 6–11 vertical rows of loreals, and no clearly demarcated sublabials. The rostral scale is in contact with the nasal scales, azygous scales in the prefrontal suture are absent, and it has 80–110 dorsal crest scales (Schwartz and Carey, 1977).

Body color varies among island populations from gray or tan to brown or dull green. In some populations, the head and neck have a vermiculated pattern. Juveniles are gray with 4–6 anterior brown crossbars that do not extend onto the sides and fade laterally, growing fainter with age.

The generic name *Cyclura* is derived from the Greek words *cyclos*, meaning circular, and *urus*, meaning tail, after the thick-ringed tail characteristic of all iguanas in the genus. The specific name *carinata* means keeled, and refers to the animal's scale type.

Natural History Notes

The Turks and Caicos iguana is one of the world's smallest rock iguanas, with males reaching up to 36.0 cm snout–vent length (SVL) and weighing up to 1.86 kg; females measure up to 29.0 cm SVL and weigh as much as 1.14 kg (G. Gerber, personal communication). Body size is

FIGURE 2.2 Turks and Caicos iguana (*Cyclura carinata carinata*).

sexually dimorphic and varies among islands, with the largest recorded specimens found on Pine Cay on the west side of the Caicos Bank (Gerber and Iverson, 2000).

The Turks and Caicos iguana inhabits numerous islands throughout the Turks and Caicos Banks, as well as Booby Cay in the Bahamas. These iguanas can reach very high population densities on islands that have low anthropogenic disturbance (Welch et al., 2004). Mitochondrial DNA variation suggests that *C. carinata* may be as old as 5 million years, and the fossil record confirms it has been present in the Turks and Caicos for at least several thousand years (Welch et al., 2004).

The Turks and Caicos iguana is most common in rocky coppice and sandy strand vegetation habitats, where it shelters in natural retreats under rocks or in self-dug burrows (Figure 2.3). The Turks and Caicos iguana is primarily herbivorous, feeding on over 60 types of plants, fruits, and flowers, although it will sometimes consume insects, mollusks, crustaceans, arachnids, lizards (including young of its own species), and carrion (Iverson, 1979; Auffenberg, 1982; G. Gerber, personal communication).

As with most rock iguanas, adult males are territorial throughout the year, and become increasingly more aggressive toward other males during the breeding season in May. Males reach sexual maturity at about 7 years or when they attain an SVL of approximately 22.0 cm and a mass of 0.33–0.48 kg. Females reach sexual maturity at 6–7 years or 18.5 cm SVL and 0.20–0.30 kg.

Females lay a single annual clutch of eggs in June, consisting of 2–11 eggs. This is the only rock iguana known to dig a nesting tunnel off the main refuge burrow (Iverson, 1979). Females defend their nests for several days to several weeks after laying, but are not territorial for the rest of the year. Wild nest temperatures have been recorded at 28–29 °C (82.4–84.2 °F). At these temperatures, incubation lasts 80–90 days. Egg size

FIGURE 2.3 Turks and Caicos iguana sharing a burrow with a shearwater chick.

averages 51.8 mm long and 31.1 mm wide and eggs weigh an average of 25.9 g. Hatchlings average 8.0 cm SVL and 14.6 g (Figure 2.4). Growth rates in juvenile Turks and Caicos iguanas have been measured at 20 mm per year until maturity. Adults grow at an average rate of 0.2–1.7 cm per year. Some individual iguanas may live 20 years and probably longer, with 90–95% adult survivorship (Gerber and Iverson, 2000).

Wasilewski and Conners (2005) have been monitoring the Booby Cay population, formerly known as the Bartsch's iguana (*C. c. bartschi*), for nearly 10 years. All size classes were

FIGURE 2.4 Hatchling Turks and Caicos iguana.

observed on every visit despite the presence of rats and goats, a strong hurricane, and human activities. Their transect surveys resulted in a population estimate of 14.5 iguanas per hectare. The diet of these lizards was similar to those in the Turks and Caicos, with the possible exception of feeding on sea grass during extremely low tides.

Although the Turks and Caicos rock iguana is one of the most abundant rock iguanas in terms of population numbers, with as many as 50,000 remaining in the archipelago, it has been reduced to less than 5% of its original range, largely due to the introduction of predators (Welch et al., 2004). A small adult body size makes this species vulnerable to introduced predators, especially dogs and cats. In the 1970s, a population of 15,000 iguanas was almost completely obliterated within five years by dogs and cats brought to Pine Cay by hotel workers (Iverson, 1978, 1979). Competitive grazing with domestic and feral livestock is a secondary threat.

In 2000, a team from the San Diego Zoo Institute for Conservation Research led by Dr. Glenn Gerber began to re-establish populations of *C. carinata* on islands lacking feral mammals by translocating iguanas from islands that were being developed or threatened by cats (Figure 2.5). To date, survival rates of translocated adult iguanas have been 98%. The average adult body size on all translocation cays was larger than that documented on the source cays, and successful reproduction has occurred on all translocation cays. In addition, the time to reach sexual maturity was reduced from 6–7 years to 1.5–2.5 years, and juvenile growth rates were four times faster than average rates on source cays (G. Gerber, personal communication).

Taxonomic Notes

The closest living relative to the Turks and Caicos iguana is Ricord's iguana, *C. ricordii* (Malone et al., 2000; Malone and Davis, 2004). Bryan et al. (2007), using mitochondrial

FIGURE 2.5 A team from the San Diego Zoo reintroduces Turks and Caicos iguanas to islands free of feral predators.

DNA, showed that the Booby Cay iguana (*C. carinata bartschi*) is not distinct, but represents another population of the Turks and Caicos iguana. Support for subspecific status is weak and it seems that geographic isolation is the only indicator of potential genetic diversification. In fact, Schwartz and Carey (1977) found that *bartschi* and *carinata* overlapped in character state distribution for all characters examined.

Conservation Status

The Turks and Caicos iguana is protected from international trade on Appendix I of the Convention on International Trade in Endangered Species of Wild Fauna and Flora (CITES) and is listed on the IUCN Red List as Critically Endangered (www.redlist.org). It is also listed as Threatened by the US States Fish and Wildlife Service. In 2003, legislation to protect the species was drafted by the Turks and Caicos government. The Booby Cay population is protected under Bahamian law, although Booby Cay is not part of the Bahamas national park system and development on nearby Mayaguana threatens the population.

JAMAICAN IGUANA (*Cyclura collei*)

Gray, 1845 Synonyms: *Cyclura lophoma*, Gosse, 1848

Description

The Jamaican iguana (*C. collei*) (Figure 2.6) is a medium-sized iguana with nasal scales that are separated from the rostral by several rows of fine granules. The prefrontal region is covered by a series of three large shields on each side. The series are separated by a double

FIGURE 2.6 Jamaican iguana (*Cyclura collei*). Reddish color is due to the ferric soils characteristic of the Hellshire Hills.

row of large, irregular scales. The canthus rostralis is formed of a group of three medium-sized keeled scales and the dorsal crest is not interrupted on either shoulder or on the rump (Barbour and Noble, 1916). It has 3–4 poorly defined vertical rows of loreals, 13–14 suboculars to the anterior border of the tympanum, 5–6 sublabials to eye center, and 1–3 postmentals (Schwartz and Carey, 1977).

Body color is green, grading into slaty blue, with lines of olive-green on the shoulder. Three triangular patches extend from the dorsal crest to the venter, with olive-brown zigzag spots. The dorsal crest scales are brighter bluish-green than the body and the lateral body surfaces are blotched with straw color, with the blotches breaking up into small groups of spots (Schwartz and Henderson, 1991). Hatchlings and young juveniles are blue-green in color with some darker mottling.

The generic name *Cyclura* is derived from the Greek words *cyclos*, meaning circular, and *urus*, meaning tail, after the thick-ringed tail characteristic of all iguanas in the genus. The specific name *collei* is Latin for hill, and refers to the hilly regions in Jamaica where the iguana was once found.

Natural History Notes

The Jamaican iguana is a medium-sized, robust species, with males reaching 42.8 cm SVL and females reaching 37.8 cm SVL (Vogel, 2000).

This species was once common throughout many parts of Jamaica, but began to decline dramatically during the second half of the nineteenth century after the introduction of the Indian mongoose (*Herpestes javanicus*) as a form of rat and snake control. At the beginning of the 1900s the iguana was thought to still survive on the Goat Islands, but was believed to be extinct by the 1940s. In 1970, a pig hunter stumbled across an iguana carcass in the Hellshire Hills, but no live animals were found. In 1990, a pig hunter's dog captured an iguana in the rugged Hellshire Hills west of Kingston, and a preliminary survey of the Hellshire Hills revealed a small remnant population of fewer than 100 iguanas and two active nesting sites. Today, iguanas remain only in the Hellshire Hills and are threatened by charcoal burning (which has degraded nearly a third of the region) and feral mammals, including mongooses, cats, dogs, and pigs. Since its rediscovery, the iguana has been studied intensively. The population has been found to be made up almost entirely of adults, as juveniles are preyed upon heavily by feral predators.

The tropical dry forest of the Hellshire Hills is made up of coarse, red ferric soil mixed with rugged limestone outcroppings. Iguanas use the natural limestone crevices and burrows they dig as shelters. They feed on leaves, fruits, and flowers of a wide variety of plant species, but they often take animal matter including snails, beetles, and spiders, as well (Vogel, 2000; R. Van Veen, personal communication).

Male iguanas are territorial throughout the year and compete with neighboring males for females during the breeding season in May and June. Repatriated females have produced offspring at 7 years of age. Mean size of reproductive females is 38.9 cm SVL and 2.27 kg. Gravid female iguanas begin digging trial holes long before egg laying commences. Females lay communally in mid-June in underground tunnel systems of burrows filled with loose soil. Tunnels range from 20 to 60 cm in length; some then turn 90° and continue at the same depth for another 30 cm. Nest depth has been recorded at 50 cm. Hatchlings emerge approximately 85–87 days later when eggs are incubated at 30 °C (86 °F). Clutch size averages 17 eggs (range 16–20). Individual eggs average 55.8 mm long, 38.9 mm wide, and weigh 39.9 g. Hatchlings average 9.5 cm SVL and 34.2 g in mass (Vogel, 1994).

Since the rediscovery of the Jamaican iguana, the species has been the subject of an intensive conservation program. Because predation by the mongoose was perceived as the biggest threat, feral mammal removal has been ongoing. Iguanas hatched in the wild have also been captured every year for headstarting and release, and many repatriated animals are breeding in the wild (Figure 2.7). With the assistance of numerous US zoos and other agencies, a headstarting and breeding center was built at the Hope Zoo in Kingston, Jamaica, soon after the rediscovery of the iguana. Since then, the species has reproduced at the zoo only twice, but the headstarting facility has been very successful, with hundreds of iguanas brought to the zoo as hatchlings being released back into the Hellshire Hills. The headstart facility carries out health screening prior to the release of specimens to determine baseline the normal physiologic values and identify potential problems related to parasites and disease that could threaten the wild population. Research also continues on the adult population, and both wild and repatriated animals have been fitted with radiotelemetry equipment to track their movements. Preliminary data reveal that females have smaller home ranges than males, and females live quite close to the known nesting areas (Wilson and Van Veen, 2007). A wealth of information has come from the Jamaican Iguana Recovery Program over the last few years thanks to the constant presence of researchers living in the Hellshire Hills among the iguanas.

Habitat protection has been the greatest challenge in conserving the Jamaican iguana. A proposal to restore an iguana population to the Goat Islands has been put forward. The Hellshire Hills has recently been afforded some tenuous protection through the Urban Development Corporation, which has been given management authority over the area.

FIGURE 2.7 Jamaican iguana nesting site in the Hellshire Hills.

FIGURE 2.8 Juvenile Jamaican iguana at the San Diego Zoo.

In 1994 and again in 1996, ex situ populations of Jamaican iguanas were established in numerous US zoos. Jamaican iguanas first bred in the United States at the Indianapolis Zoo in 2006 (Figure 2.8).

Taxonomic Notes

Malone et al. (2000) found that the Jamaican iguana is one of the most genetically unique *Cyclura* species and that it is most closely related to *C. lewisi, C. rileyi, C. cornuta,* and *C. cychlura*.

Conservation Status

The Jamaican iguana is protected from international trade on Appendix I of CITES and is listed on the IUCN Red List as Critically Endangered (www.redlist.org). It is also listed as Endangered by the US Fish and Wildlife Service.

RHINOCEROS IGUANA (*Cyclura cornuta cornuta*)

Bonnaterre, 1789 Synonyms: *Lacerta cornuta*, Bonnaterre, 1789; *Iguana cornuta*, Lacèpede, 1789; *Metopoceros cornutus*, Wagler, 1830; *Hypsilophus cornutus*, Fitzinger, 1843; *Cyclura cornuta cornuta*, Cope, 1886

Description

The Rhinoceros iguana (*C. c. cornuta*) (Figure 2.9) is a large rock iguana characterized by 3–5 bony tubercles on the snout that resemble horns, a combination of two rows of scales between the prefrontal shields and the frontal scale, three rows of scales between the supra-orbital semicircles and interparietal, eight supralabials to eye center, and 12 sublabials to eye center (Schwartz and Carey, 1977).

Adults are dark brown to dark gray or black without pattern and with the venter less heavily pigmented than the dorsum. Juveniles are similar in appearance to adults, but have nine pale crossbars that fade soon after hatching.

The generic name *Cyclura* is derived from the Greek words *cyclos,* meaning circular, and *urus,* meaning tail, after the thick-ringed tail characteristic of all iguanas in the genus. The Rhinoceros iguana's specific name, *cornuta,* derives from the horned projections on the snouts of males of the species.

Natural History Notes

The Rhinoceros iguana is a large heavy-bodied iguana, with males reaching 56.0 cm SVL and up to 10 kg in mass; females reach 51.0 cm SVL (Ottenwalder, 2000a). In addition to larger overall body size, males also have a larger dorsal crest and head, and larger femoral pores than females.

FIGURE 2.9 Adult female Rhinoceros iguana (*Cyclura cornuta cornuta*).

Rhinoceros iguanas are still widely distributed throughout Hispaniola, including most offshore islands. Their current geographic range is fragmented as a result of their association with xeric regions of lower human population density. Important threats to this species are habitat destruction stemming from human activities such as charcoal production, agriculture, mining, and livestock grazing. Feral dogs, cats, mongooses, and pigs also threaten iguanas, as well as illegal hunting by humans. Most iguana concentrations are found along the southern side of Hispaniola, with the highest numbers in south-southwestern Dominican Republic. Populations are probably only stable on Isla Beata and the extreme of the Barahona Peninsula inside Parque Nacional Jaragua (Ottenwalder, 2000a). The species is also present on Isla Cabritos, a small island in the middle of Lago Enriquillo within Parque Nacional Lago Enriquillo. In this location, the species is sympatric with Ricord's iguana (*C. ricordii*). Rhinoceros iguanas are most abundant in xeric, rocky habitats with an elevational range of −35 meters (on Isla Cabritos) to 400 meters. They are found in a variety of habitats, including thorn scrub woodlands, dry forests, and semideciduous to subtropical moist forests (Ottenwalder, 2000a). They shelter in rock crevices or self-dug burrows, as well as hollow tree trunks and caves. The Rhinoceros iguana is primarily herbivorous, but like most rock iguanas, probably feeds on some animal matter, especially caterpillars and pupae.

Adult male Rhinoceros iguanas defend territories containing retreats that are attractive to females (Ottenwalder, 2000a) (Figure 2.10). Sexual maturity is reached in two to three years, and breeding takes place at the beginning of the rainy season in May and lasts through June. Up to 34 eggs are laid in June in a tunnel approximately one meter long with a chamber at the end just large enough for the female to turn around. Eggs can be as large as 82.6 mm long, by 50.8 mm wide, and weigh 75–80 g (Boylan, 1984). Females often guard the nest site for several days after laying. Hatchlings emerge from the nest after approximately 85 days of incubation at 31 °C (87.8 °F). Hatchlings measure an average of 10.4 cm SVL and weigh an average of 51 g (Figure 2.11).

FIGURE 2.10 Adult male Rhinoceros iguanas fighting over territory.

FIGURE 2.11 Hatchling Rhinoceros iguana.

Observations on population and habitat trends recorded since the 1970s estimate that 10,000 to less than 17,000 Rhinoceros iguanas remain in the wild. Iguana densities are low in the majority of areas where they presently occur and appear to be declining. Local extirpations are known from both the Dominican Republic and Haiti, and only ten subpopulations of Rhinoceros iguanas are known from the Dominican Republic. In Haiti, ten or fewer threatened subpopulations may still exist, but accurate information concerning population estimates is lacking. In the Dominican Republic, most Rhinoceros iguana populations are either fully or partially protected in national parks and reserves, whereas in Haiti no protected areas have been established (Ottenwalder, 2000a).

Rhinoceros iguanas are the most common rock iguana species in captivity, both in zoos and private collections. Breeding and release programs managed by ZooDom in Santo Domingo have been used to supplement wild iguana populations for many years. Current distributions and population estimates are being assessed by Grupo Jaragua and other Dominican conservation groups.

Taxonomic Notes

The subspecies of the Rhinoceros iguana (*cornuta*, *stejnegeri*, and *onchiopsis*) are most closely related to the Turks and Caicos iguana (*C. c. carinata*) and the Ricord's iguana (*C. ricordii*) (Malone et al., 2000).

Conservation Status

The rhinoceros iguana is protected from international trade on Appendix I of CITES and is listed on the IUCN Red List as Vulnerable (www.redlist.org). It is also listed as Threatened by the US Fish and Wildlife Service.

MONA ISLAND IGUANA (*Cyclura cornuta stejnegeri*)

Barbour, 1937 Synonyms: *Cyclura stejnegeri*, Barbour and Noble, 1916

Description

The Mona Island iguana (*C. c. stejnegeri*) (Figure 2.12) is a large iguana that differs from the Rhinoceros iguana (*C. c. cornuta*) in that it has three scale rows between the prefrontal shields and the frontal scale, 12–14 subocular scales to the tympanum, eight sublabial scales to the center of the eye, and rostral scale and nasal scales often in contact. *C. c. stejnegeri* also has a higher incidence of double or triple rows of femoral pores (Schwartz and Carey, 1977).

Body color is a uniform gray to olive with slight brown or blue colorations in adults. Juveniles have a faint pattern of eight pale crossbars on the sides, separated by gray areas that may form series of blotches. This pattern fades with age.

The generic name *Cyclura* is derived from the Greek words *cyclos*, meaning circular, and *urus*, meaning tail, after the thick-ringed tail characteristic of all iguanas in the genus. The Mona Island iguana's specific name, *cornuta*, derives from the horned projections on the

FIGURE 2.12 Mona Island iguana (*Cyclura cornuta stejnegeri*).

snouts of males of the species. The subspecific name, *stejnegeri*, was chosen to honor naturalist Leonhard Hess Stejneger.

Natural History Notes

The Mona Island iguana is a robust species, with males reaching up to 6.1 kg in mass and 51.7 cm SVL. Females are generally smaller, weighing up to 4.7 kg and measuring up to 47.5 cm SVL (Wiewandt and Garcia, 2000).

The Mona Island iguana is found only on Isla Mona, a rocky island situated between Puerto Rico and Hispaniola. Iguanas are found on the majority of the island, and use rocks, caves, and deep burrows as retreats. Deep sinkholes are used by both males and females, and males often defend sinkholes that are used by females that live within their home ranges. Mona Island iguanas are primarily herbivorous and are particularly attracted to fruits that fall from native trees. Iguanas consume 71 of the 400 plants that are found on Mona (Wiewandt, 1977). Iguanas have also been observed feeding on introduced breadfruit (*Artocarpus altilis*) on Mona (J. Lemm, personal observation). Animal matter is eaten when available, and caterpillars of sphingid moths are relished by Mona Island iguanas (Powell, 2000a).

Average home range size of adult males has been recorded as 0.60 hectares, and average home range size of adult females is 0.28 hectares (Perez-Buitrago et al., 2007). Overlap of male home ranges is very low, as is overlap of female home ranges, but overlap between the home ranges of males and females is significant.

Females reach maturity at 38.0 cm SVL and about 2.0 kg, requiring approximately 6–7 years to reach this size (Figure 2.13). Mating takes place in June with oviposition in July. Because most of Mona is comprised of rock, females either lay in sinkholes or travel to the coast to nest in sandy areas. Clutch size averages 12 eggs, and eggs average 81.4 mm in length by 51.1 mm in width and weigh an average of 107 g. Females often fight over prime nesting

FIGURE 2.13 Female Mona Island iguana (*C. c. stejnegeri*).

areas and defend their nests aggressively. Incubation lasts an average of 83 days at temperatures ranging from 30 to 33 °C (86–91.4 °F). Hatchling Mona Island iguanas average 11.9 cm SVL and 74.0 g (Wiewandt and Garcia, 2000). Hatchlings disperse up to 5,080 meters from the hatching site, and once settled, spend more than 60% of the time in trees or other elevated areas. Hatchling home range size averages 297 square meters and hatchling survival has been documented at 22% (Perez-Buitrago and Sabat, 2007).

The population of Mona Island iguanas has remained relatively stable for the past 25 years. The most recent population census estimates a population more than 5,000 adult iguanas, with a density of 0.96 iguanas per hectare (Perez-Buitrago and Sabat, 2000). Iguana abundance varies with habitat type and the presence of sinkholes that are used as refugia. Low densities and the fact that juvenile recruitment is depressed due to predation by feral animals suggest a senescent population on Mona. The greatest threat to Mona Island iguanas is feral cats that prey on juveniles. Feral pigs feed heavily on iguana eggs and in some areas have been known to predate up to 100% of nests (Wiewandt, 1977). Introduced goats also feed heavily on iguana food plants.

Since 1999, the Puerto Rico Department of Natural Resources and the Environment has been managing the Mona Island iguana population. Headstart-release and feral animal removal programs, as well as habitat management, have been implemented with success.

Taxonomic Notes

The Mona Island iguana is most closely related to the Rhinoceros iguana (*C. c cornuta*). As a species, *C. cornuta* is most closely related to Ricord's iguana (*C. ricordii*) and the Turks and Caicos iguana (*C. c. carinata*) (Malone and Davis, 2004). Some consider the Mona Island iguana to be a separate species based on scale counts (Schwartz and Carey, 1977; Powell and Glor, 2000).

Conservation Status

The Mona Island iguana is protected from international trade on Appendix I of CITES and is listed on the IUCN Red List as Endangered (www.redlist.org). It is also listed as Endangered by the US Fish and Wildlife Service.

ANDROS ISLAND IGUANA (*Cyclura cychlura cychlura*)

Cuvier, 1829 Synonyms: *Iguana cyclura*, Cuvier, 1829; *Cyclura baeolopha*, Cope, 1861; *Cyclura cychlura cychlura*, Schwartz and Thomas, 1975

Description

The Andros Island iguana (*C. c. cychlura*) (Figure 2.14) is a large rock iguana that is one of three weakly-defined *C. cychlura* subspecies present in the northern Bahamas (Schwartz and Carey, 1977). The species is characterized by 3–7 scale rows between the frontal and interparietal scales. Azygous scales are often present between the four prefrontal scales. The posterior

FIGURE 2.14 Adult male Andros iguana (*Cyclura cychlura cychlura*).

canthal scale is enlarged and elongate and the posterior pair of prefrontal scales is enlarged in large males. There are 6–12 scales bordering the frontal scale and the first prefrontal scale is almost always in contact with the precanthal scale. The species has 4–9 vertical rows of loreals, 9–17 suboculars to the anterior border of the tympanum, and the rostral scale always in contact with the nasal scales. Dorsal crest scales average 96 (range 67–118) (Schwartz and Carey, 1977; Iverson et al., 2006a). The subspecies is characterized by having frontal scales separated from posterior prefrontals by two scale rows. There are usually no scales in the suture between the four prefrontal scales. Two small scales are present between the posterior canthal scale and the precanthal scales and accessory scales are usually present between the infralabial and sublabial scale rows. There is usually only one postmental scale present (Schwartz and Carey, 1977; Schwartz and Henderson, 1991; Iverson et al., 2006a). Dorsal crest scales average 96.5 (range 84–108).

Body color of *C. c. cychlura* varies from dark gray to black, with yellowish-green or orange-tinged scales on the legs, dorsal crest, and head. The yellow often changes to orange-red with maturity (Auffenberg, 1976). Juveniles are often gray with 5–7 pale chevrons present dorsally, alternating with darker chevrons that fade with age (Iverson et al., 2006a).

The generic name *Cyclura* is derived from the Greek words *cyclos*, meaning circular, and *urus*, meaning tail, after the thick-ringed tail characteristic of all iguanas in the genus. The same is true for the specific/subspecific name *cychlura*. The different spellings for these two names was intentional by the respective authors (Iverson et al., 2006a).

Natural History Notes

The Andros Island iguana is among the largest of the rock iguanas. Males may reach lengths of more than 127 cm from head to tail and often weigh more than 9.0 kg (C. Knapp, personal communication). Females reach 57.0 cm SVL and 115 cm total length (Iverson et al., 2006a). Body size is sexually dimorphic.

The Andros iguana is the only Bahamian iguana not restricted to small cays (Knapp, 2005a). Due to the difficulty of surveying in rugged terrain and shallow water channels, the true distribution and number of iguanas that remain is unknown. Andros Island lies on the western edge of the Great Bahama Bank and is a composite of three large islands (North Andros, Mangrove Cay, and South Andros) and numerous smaller islands separated by tidal channels. The preferred habitat of this iguana is under the open canopy of the pine barrens, where limestone karst rock provides adequate shelter. Shrublands, mangrove, and closed canopy pine barrens are also used by iguanas (Knapp, 2005a). In common with many species of wild rock iguana, Andros iguanas can also be found close to areas of human activity (Figure 2.15). Andros iguanas are herbivorous, and feed on the fruits, flowers, and leaves of a variety of plants.

Like the other rock iguanas, the Andros iguana is territorial and home ranges often change through the year. Knapp (2005a) found that home ranges for males and females expand in the breeding season and that individual animals use different areas throughout the year. Home ranges increase in the breeding season for males, but not for females, and home ranges overlap in the breeding season for both males and females, which increases the number of

FIGURE 2.15 Wild rock iguanas often become accustomed to humans. This very large pair of Andros iguanas was using this construction material as a basking area.

aggressive encounters between males. Average male home range size (95% kernel) is 11.53 hectares, while average female home range size is 2.19 hectares.

Female Andros iguanas can reproduce successfully at 31.0 cm SVL and 1,279 g (Knapp, 2005a; Knapp et al., 2006). It is unknown when males reach sexual maturity, but like other rock iguanas, both males and females can probably reproduce from anywhere between three and five years of age. Breeding takes place in April and May and females generally lay eggs in May and June (Knapp, 2005a). The majority of clutches are deposited in open pine habitat. The Andros iguana is unique among iguanas in that nearly all females deposit their eggs in active termite mounds (Figure 2.16). Female iguanas dig at the base of the termite mound and angle the tunnel upwards. Digging may last up to five days (Knapp, 2005a). Knapp (2005a) also reports that iguanas lay their eggs outside of the tunnel entrance and push eggs into the termite mound with their front limbs and snout. Females use their snouts and front limbs to pack the nest tunnel, and then scratch dirt, pine needles, and other materials over the entrance hole. Females guard their nests for up to six weeks. As many as 19 eggs are laid in a single annual clutch (Buckner and Blair, 2000a). Eggs incubate for from 72 to 82 days at a temperature of 30.8 °C (Knapp, 2005a). Eggs average 70.7 mm long by 40.3 mm wide and weigh an average of 59.8 g. Hatchlings average 9.6 cm SVL and weigh 37 g (Knapp et al., 2006).

FIGURE 2.16 Andros iguanas are the only rock iguana known to lay eggs in termite mounds. Females often defend the nests aggressively. *Photo by Chuck Knapp.*

Taxonomic Notes

The Andros Island iguana is most closely related to the other *C. cychlura* subspecies, the Exuma Island iguana (*C. c. figginsi*) and the Allen Cays iguana (*C. c. inornata*). Because the Andros population was fragmented from the Bahama Bank after the Pleistocene glacial maximum sea level, it is phylogenetically distinct from the Exuma Cay populations. As a species, *C. cychlura* is most closely related to the *C. nubila* (Malone et al., 2000).

Conservation Status

The Andros Island iguana is protected from international trade on Appendix I of CITES and is listed on the IUCN Red List as Vulnerable (www.redlist.org). It is also listed as Endangered by the US Fish and Wildlife Service. In addition, all Bahamian rock iguanas are protected under the Bahamas Wild Animals Protection Act of 1968.

EXUMA ISLAND IGUANA (*Cyclura cychlura figginsi*)

Barbour, 1923 Synonyms: *Cychlura figginsi*, Barbour, 1923

Description

The Exuma Island iguana (*C. c. figginsi*) (Figure 2.17) is the smallest of the three weakly-defined *C. cychlura* subspecies from the northern Bahamas (Schwartz and Carey, 1977).

FIGURE 2.17 Exuma Island iguana (*Cyclura cychlura figginsi*). *Photo by Chuck Knapp.*

This subspecies of rock iguana is characterized by three rows of scales between the three prefrontal shields and frontal scale, one azygous scale in the prefrontal suture, four scales between the anterior canthal scales, 10 superciliary scales, and one postmental scale (Schwartz and Carey, 1977). Barbour (1923) describes *figginsi* as having tiny supranasals compared to the other subspecies. Total dorsal crest scales range from 67 to 118 (mean 92.7), and there are 41–66 (mean 49.4) scales around the tail in the fifth caudal verticil (Schwartz and Carey, 1977).

Body color is variable between populations. Animals from Bitter Guana and Gaulin Cays are dull gray-brown above with white or light red dorsal crest scales. The head scales are usually tinged in black with orange on the snout and infralabials (Schwartz and Carey, 1977; Knapp, 2000a). Adults from Guana Cay are dull black with diffuse pale white ventral and gular coloration. Upper labial, temporal, parietal, nuchal, and ocular scales may be light blue in color, while dorsal crest scales are either scarlet or gray tinged with red (Knapp, 2000a). Juvenile iguanas have seven black body bands that become slightly diagonal laterally, alternating with pale gray bands. Bands are mottled with small pale dots and fade with age (Schwartz and Carey, 1977) (Figure 2.18).

The generic name *Cyclura* is derived from the Greek words *cyclos*, meaning circular, and *urus*, meaning tail, after the thick-ringed tail characteristic of all iguanas in the genus. The same is true for the specific name *cychlura*. The different spellings for these two names was intentional by the respective authors (Iverson et al., 2006a). The subspecific name *figginsi* honors J. D. Figgins, former director of the Colorado Museum of Natural History.

Natural History Notes

The Exuma Island iguana is regarded as the smallest of the three subspecies of *C. cychlura*, although body size varies among islands. Knapp (2000a) found that adult male animals range

FIGURE 2.18 Head detail of juvenile Exuma Island iguana.

from 470 mm SVL and 3.25 kg to 542 SVL and 8.15 kg. Body size is sexually dimorphic and the largest animals are known from the northern extent of the range where a new, large-bodied population of iguanas was discovered in 1997 (Knapp, 2000a) (Figure 2.19).

Exuma Island iguanas are found on at least seven small cays throughout the central and southern Exuma Island chain in the northern Bahamas. They can often be found in high densities in a variety of habitats ranging from sandy beaches and limestone areas devoid of vegetation to vegetated, rocky areas and areas with tall trees (Wilcox et al., 1973; Knapp, 2000a). Exuma Island iguanas use limestone crevices and sand burrows that they dig themselves for nightly retreats. These iguanas are herbivorous and feed on the leaves, flowers, and berries of over 100 types of plants. They have been documented feeding both on the ground and in trees (Windrow, 1977; Knapp, 2000a). Coenen (1995) reports that during the summer months, iguanas spend a great deal of time foraging for droppings of White-crowned pigeons (*Columba leucocephala*) and Zenaida doves (*Zenaida aurita*). More recently, ecotourism has had a negative impact on iguana health on islands where tourists have altered natural diets by feeding iguanas. Preliminary data from Hines (2007) have shown that 30% of iguana scats from the main feeding beaches are made up primarily of grapes and sand, resulting in a high incidence of diarrhea.

Unlike other rock iguanas, Exuma Island iguanas do not display territorial or hierarchical behavior throughout the majority of their range, even in areas with high iguana densities (Carey, 1976; Knapp, 2000a). Carey (1976) believes that this lack of social structure allows the population to remain dense under conditions of limited resources. Nonetheless, these iguanas occasionally demonstrate aggression, usually when personal space is violated, or over preferred food items. Iguanas display to one another with headbobs and sometimes chase one another, with the largest animal emerging victorious (Windrow, 1977; Coenen, 1995). Captive animals at the San Diego Zoo exhibit some aggressive behavior during the warmer months of the year, primarily during the breeding season. Males and females chase one another, although there does not seem to be a single dominant animal (K. Morgan, personal communication).

FIGURE 2.19 Adult Exuma Island iguana. *Photo by Chuck Knapp.*

Exuma Island iguanas probably breed in May and lay eggs in June. It is unknown at what size or age male iguanas become mature, but mean size of reproductive females is 28.5 cm SVL and 960 g. Coenen (1995) reports that a small female iguana entered another female's nest and begin to dig. The smaller intruder weighed 539 g and measured 61 cm SVL. Only two nests have ever been found in the wild, both in early June (Coenen, 1995). Both nests were roughly two feet long and 3–5 inches below the surface. Each contained only three eggs that measured an average of 85 mm long, 40 mm wide, and 51 g (Coenen, 1995). Hatchlings have been found in late June and early July (Figure 2.20). A hatchling captured in July weighed 35.6 g and measured 11 cm SVL and had a tail length of 28 cm (Coenen, 1995).

Taxonomic Notes

The Exuma Island iguana is most closely related to the other *C. cychlura* subspecies, the Andros Island iguana (*C. c. cychlura*) and the Allen Cays iguana (*C. c. inornata*) (Malone et al., 2003). However, data show that *figginsi* and *inornata* are very similar and may not represent separate taxa (Malone et al., 2003; Malone and Davis, 2004). As a species, *C. cychlura* is most closely related to the *C. nubila* (Malone et al., 2000).

Conservation Status

The Exuma Island iguana is protected from international trade on Appendix I of CITES and is listed on the IUCN Red List as Endangered (www.redlist.org). It is also listed as Critically Endangered by the US Fish and Wildlife Service. In addition, all Bahamian rock iguanas are protected under the Bahamas Wild Animals Protection Act of 1968.

FIGURE 2.20 Hatchling Exuma Island iguana. *Photo by Joe Wasilewski.*

ALLEN CAYS IGUANA (*Cyclura cychlura inornata*)

Cuvier, 1829 Synonyms: *Cyclura inornata*, Barbour and Noble, 1916

Description

The Allen Cays iguana (*C. c. inornata*) (Figure 2.21) is a large subspecies of rock iguana found in the northern Bahamas. This subspecies of *C. cychlura* has four rows of scales between the prefrontal shields and frontal scale, one azygous scale in the prefrontal suture, and six scales between the anterior canthal scales. There are nine superciliaries, 14 suboculars to the anterior border of the tympanum, and two postmental scales. Total dorsal crest scales number 83–110 (mean 104.5), and there are 40–55 (mean 45.9) scales around the tail in the fifth caudal verticil (Schwartz and Carey, 1977). This subspecies lacks the horn-like frontal or prefrontal scales, and the rostral scale is in contact with the nasal scales (Iverson, 2000). The Allen Cays iguana is pigmented gray–black with cream, pink, or orange mottling on the dorsum. The pink or orange pigment is obvious on the lower labial scales, the preauricular scales, and the enlarged mid-dorsal scale row (Iverson, 2000). The rostrum is black to brownish-black (Iverson et. al, 2006a). Juveniles are often gray with 5–7 pale chevrons present dorsally, alternating with darker chevrons that fade with age (Iverson et al., 2006a).

The generic name *Cyclura* is derived from the Greek words *cyclos*, meaning circular, and *urus*, meaning tail, after the thick-ringed tail characteristic of all iguanas in the genus. The same is true for the specific name *cychlura*. The different spellings for these two names was intentional by the respective authors (Iverson et al., 2006a). The subspecific name

FIGURE 2.21 Allen Cays iguana (*Cyclura cychlura inornata*). *Photo by John Iverson.*

inornata means undecorated, or not beautiful, and probably refers to the dull-colored female the describing author had available (Iverson et al., 2006a).

Natural History Notes

The Allen Cays iguana is large in size; adult males can reach 62 cm SVL and 10.5 kg in mass and females reach 57 cm SVL and 6.9 kg (J. Iverson, personal communication). Although males grow larger than females, they are nearly identical in appearance, which is unique among rock iguanas.

The Allen Cays iguana is the most northern member of the genus and was nearly extirpated in the 1900s due to heavy hunting pressure by humans for food. By 1970, about 150 lizards remained on Leaf and U Cays. Today it is estimated that approximately 500 adult animals exist in the wild.

This subspecies was once known only from Leaf Cay and U Cay in the Northern Exuma Island chain in the Bahamas, but recent unauthorized translocations by unknown individuals have apparently placed Allen Cays iguanas on nine or more islands. Some iguanas have been found as far as 6 km north of their known distribution (Hines and Iverson, 2006). Allen Cays iguanas are found in all habitats on Leaf and U Cays, including bare limestone rock. The iguanas seem to be able to survive in any habitat as long as food is present. However, lack of breeding on some cays may be due to insufficient areas of exposed sand for nesting (Iverson, 2000). Like all rock iguanas, Allen Cays iguanas are diurnally active, spending the nights in burrows they have dug, in rocks, or in other natural retreats. They feed primarily on vegetation, including as fruits, flowers, and leaves, but are opportunistic carnivores as well, as evidenced by crab claws found in their feces (Iverson, 2000). Humans regularly feed the iguanas, which may be contributing to their decline on some islands (Figure 2.22). On Leaf Cay, where the greatest population of Allen Cays iguanas is found, over 100 tourists visit the island daily to feed the iguanas through eco-tour operators. Hines (2007) collected iguana fecal samples from the primary iguana feeding beach on Leaf Cay and found that about 30% contained high concentrations of grapes and sand. The iguana population had a high incidence of diarrhea and fecal samples from these animals appeared as cement-like tubes.

During the non-breeding season, Allen Cays iguanas appear to have dominance hierarchies rather than defended territories. This is probably partly the result of a breakdown in social structure due to feeding by tourists (Iverson, 2000). However, in studying an introduced population on Alligator Cay, Knapp (2000b) found that food provisioning is not the only factor responsible for the lack of consistent territorial behavior in this subspecies. He believes that small island habitation and high population density are responsible for the lack of territorial behavior, as evidenced by lack of tail break frequency and home range overlap. Although only nine animals were studied, Knapp found that home range size in male *allen cays iguanas* averaged 0.30 hectares and home range size in females averaged 0.02 hectares. Percent home range overlap averaged 86.8% for all monitored males, but only 24% for the four larger founder males (over 39.0 cm SVL). There was no home range overlap in the two adult females studied. There was also a correlation between percent home range overlap in mature males and body size, with home range overlap inversely proportional to mature male SVL and mass.

FIGURE 2.22 Tourists often feed iguanas on the Allen Cays. This action may actually be detrimental to the health of the iguanas and to iguana populations. *Photo by John Iverson.*

Female Allen Cays iguanas reach sexual maturity at 26–27 cm SVL, 1,339 g in mass, and about 12 years of age (Iverson et al., 2004). The oldest known females to successfully nest are at least 40 years old, with some estimated to be 43 years of age. One in three females nest each season, with only the largest females nesting every year. About one in five smaller females (26–30 cm SVL) nest each year. Mating occurs in mid-May (Figure 2.23) and females travel an average of 88 meters from home range to nesting areas. Oviposition

FIGURE 2.23 Copulating Allen Cays iguanas. *Photo by John Iverson.*

occurs in June, roughly 35 days after copulation. Nest burrows average 149 cm in total length with an average depth to the bottom of the egg chamber of 27.7 cm. Burrow structure varies tremendously, but usually egg chambers are situated off the main burrow in a J-shaped configuration. Burrows are not associated with any particular type of plant, and tend to be in open areas with direct sun exposure. Densiometer readings of nests (a measure of percent shadiness) average 22% and nests in shadier areas tend to be more shallow than those in sunnier areas. Almost half of all nests are situated in the exact same spot each year. Clutch size averages 4.6 eggs (range 1–10) and eggs weigh an average of 49.1 g. Egg length averages 67.2 mm and egg width averages 34.9 mm. Hatching occurs after an incubation period of 80–85 days at an average temperature of 31.4 °C. Hatchlings average 34.7 g in mass and 9.5 cm SVL. Survivorship from egg to hatchling emergence has been recorded at 75.2% (Iverson et al., 2004). Hatchling survival of Allen Cays iguanas is higher than has been documented in any other species (Iverson, 2007). Of 16 juveniles marked at 6 months of age, nine were alive at 12.5 years. Of four juveniles marked at 1.5 years of age, three were alive 10 years later.

Average growth rates of Allen Cays iguanas are over 20 mm SVL per year during the first year, declining to 15 mm per year by age 5.5 (at 206 mm SVL) (Iverson, 2000). Adult males grow significantly faster than adult females. Adult males grow an average of 1.764 cm SVL per year whereas adult females grow an average of 1.139 cm SVL per year (Iverson and Mamula, 1989). In their landmark ongoing study of the Allen Cays iguana, which has been running for more than 30 years, Iverson et al. (2006b) found that in areas with less human visitation iguana survival is higher than in areas where people visit daily, and females (which are less bold than males) have higher survivorship. Populations on both Leaf and U Cays have more than doubled over 25 years, but population growth rates have declined to near zero in recent years, suggesting populations may be at or near carrying capacity. In addition, Smith and Iverson (2006) found that the sex ratios of iguanas in both populations have changed from strongly male-biased initially to an even ratio in recent years. It is believed that this is due to the recovery of the populations from intense harvesting, particularly of females, and also to the removal of males by poachers and tourists.

Taxonomic Notes

The Allen Cays iguana is most closely related to the other *C. cychlura* subspecies: the Andros Island iguana (*C. c. cychlura*) and the Exuma Island iguana (*C. c. figginsi*) (Malone et al., 2003). However, data show that *figginsi* and *inornata* are very similar and may not represent separate taxa (Malone et al., 2003; Malone and Davis, 2004). As a species, *C. cychlura* is most closely related to *C. nubila* (Malone et al., 2000).

Conservation Status

The Allen Cays iguana is protected from international trade on Appendix I of CITES and is listed on the IUCN Red List as Endangered (www.redlist.org). It is also listed as Endangered by the US Fish and Wildlife Service. In addition, all Bahamian rock iguanas are protected under the Bahamas Wild Animals Protection Act of 1968.

GRAND CAYMAN BLUE IGUANA (*Cyclura lewisi*)

Grant, 1940 Synonyms: *Cyclura nubila lewisi*, Grant, 1940; *Cyclura macleayi lewisi*, Grant, 1940; *Cyclura lewisi*, Burton, 2004a

Description

The Grand Cayman Blue iguana (*C. lewisi*) (Figure 2.24) is a large iguana that is characterized primarily by its powder-blue coloration. It is often difficult to distinguish this species from *C. nubila* by scalation alone. Schwartz and Carey (1977) distinguish what was then known as *C. n. lewisi* from other *nubila* in that *lewisi* has four scales between the prefrontal shields and the frontal scale, five scales between the anteriormost canthals, the first prefrontal very rarely in contact with the precanthal, vertical loreal rows modally 7–8, subocular scales 14–16, and four scales mid-dorsally in the fifth caudal verticil. Total dorsal crest scales number 72–80. According to Burton (2004a), Schwartz and Carey identified no uniquely diagnostic scale characters. Burton notes that scalation of the head can sometimes distinguish *C. lewisi* from *C. n. nubila* and *C. n. caymanensis; lewisi* usually has five enlarged aurical spines forming a continuous arc immediately anterior to the auricle, the most dorsal also forming the posterior terminus of the subocular scale row. Many *lewisi* have an extra pair of enlarged prefrontals immediately posterior to the first prefrontals. These second prefrontals are always smaller than the first pair and are variable in size. A third row of moderately enlarged prefrontals also occurs in *lewisi,* comparable to what appears as the second prefrontal row in *nubila* and *caymanensis*.

FIGURE 2.24 Grand Cayman Blue iguana (*Cyclura lewisi*).

Body color of adult Grand Cayman Blue iguanas is bright blue except for the feet, distal tail, and any remaining chevrons, which remain black. Hatchlings have a gray base color and alternating gray and mottled cream chevrons (Figure 2.25). The feet are patterned with ocellations and the tail is usually banded and/or streaked. Animals usually begin to turn blue and lose their juvenile coloration within a year of age (Burton, 2004a).

The generic name *Cyclura* is derived from the Greek words *cyclos*, meaning circular, and *urus*, meaning tail, after the thick-ringed tail characteristic of all iguanas in the genus. The specific name *lewisi* refers to Bernard Lewis of the Institute of Jamaica, who captured the first specimens in 1938.

Natural History Notes

The Grand Cayman Blue iguana is a large iguana; males measure up to 515 mm SVL and can reach at least 10 kg in mass (Schwartz and Carey, 1977; Burton, 2000). Females measure up to at least 410 mm SVL. Found only on the island of Grand Cayman, Blue iguana populations have been scarce ever since their original collection in 1938. Although fossil evidence indicates this species was formerly widespread in dry and coastal habitats throughout Grand Cayman, it is currently restricted to the eastern interior of the island, with extremely rare occurrences south of the Queen's Highway. The range of the Blue iguana has contracted significantly over the past 25 years, with many once-populated sites now showing no signs of wild iguanas. The most recent comprehensive survey in 2002 indicated a total population in the range 10–25 individuals. By 2005, the unmanaged wild population was considered to

FIGURE 2.25 Hatchling Grand Cayman Blue iguanas at the San Diego Zoo Institute for Conservation Research.

be functionally extinct (Burton, 2010), making the Blue iguana the most endangered lizard in the world.

The Grand Cayman Blue iguana is similar to other rock iguanas in that it spends most of its time on the ground and takes refuge in burrows and rock holes. Hatchlings and juveniles often use trees for cover. Although they were once found throughout the drier areas and coast of Grand Cayman, Blue iguanas are now primarily found inland in natural xerophytic shrubland (Burton, 2010). The diet of the Blue iguana primarily consists of plants. The leaves, fruits, and flowers of over 100 species are consumed, as are fungi, insect larvae, slugs, dead birds, and crabs (Burton, 2010).

Because Blue iguanas survive at such low densities, it is difficult to establish just how territorial this species is. It is presumed that males become more aggressive and territorial during the breeding season. Animals at the Queen Elizabeth II Botanic Park on Grand Cayman were fitted with radiotransmitters and tracked in 2004. Female Blue iguanas had home ranges of 0.6 acres and males had home ranges of 1.4 acres (Burton, 2006). Home ranges overlapped, and population density at the park was 4–5 animals per hectare. Interestingly, iguanas at the Queen Elizabeth II Botanic Park also preferred modified habitat to natural habitat, and used artificial retreats more often than natural ones. In addition, many females nested in artificial substrates (Goodman et al., 2005).

Mating generally occurs in April and May. Both males and females can reach sexual maturity in 2–3 years or at roughly 20 cm SVL (F. Burton, personal communication). Nesting takes place in June and July. Up to 22 eggs are laid in burrows that can be up to 2 meters long. The base of the nest chamber is usually 30 cm below surface; the access tunnel is closed with compacted soil and the earth around it is smoothed to disguise the tunnel. As with most rock iguanas, the nest is often guarded by the female for up to two weeks or more. Incubation temperatures in the wild have been recorded at 32 °C (89.6 °F). Incubation lasts from 66 to 80 days. Eggs average 48 mm wide by 68 mm long and weigh an average of 85 g. Hatchlings average 94.3 mm SVL and 44.0 g in mass (Goodman and Burton, 2005; R. Goodman, personal communication), and may grow 3–5 cm per year until they reach maturity. Like other rock iguanas, Blue iguanas are very long-lived. The known longevity record for this species is 69 years for a male, Godzilla, who spent 54 of his 69 years in captivity (Adams, 2004).

In 1990, a breeding program was established on Grand Cayman through the National Trust for the Cayman Islands that is now managed under the auspices of the Blue Iguana Recovery Program (Figure 2.26). In addition, numerous US zoos started to breed Blue iguanas for eventual repatriation to Grand Cayman. Unfortunately, the majority of the US founder stock were found to be hybrids of *C. lewisi* and *C. nubila caymanensis*. After many years of genetic research, pure founder animals have now begun to breed successfully in US zoos, where they are to remain as a genetic reservoir in case of catastrophic loss on Grand Cayman.

The Blue Iguana Recovery Program has been very successful at reproducing and headstarting iguanas on Grand Cayman, and progeny are now being released back into the wild (Burton, 2010). After a series of releases of captive-reared young, restored free-roaming subpopulations in two regions of Grand Cayman, the Queen Elizabeth II Botanic Park and the Salina Reserve, now total over 300 individuals. The restored subpopulation in the Queen Elizabeth II Botanic Park has been breeding since 2001,

FIGURE 2.26 Adult Grand Cayman Blue iguana breeding enclosures at the Blue Iguana Recovery Program facility in Grand Cayman.

and the subpopulation in the Salina Reserve began breeding in 2006. Individuals will be translocated between subpopulations to maintain gene flow such that the entire population remains a single genetic management unit. It is estimated that at least 1,000 animals must be released into the wild in order to achieve a viable breeding population. Once the wild population reaches carrying capacity, headstarting and release will no longer be necessary.

Taxonomic Notes

The closest relatives of the Grand Cayman Blue iguana are the Sister Isles iguana (*C. nubila caymanensis*), the Cuban iguana (*C. n. nubila*), and the northern Bahamian rock iguanas (*C. cychlura*). According to Malone et al. (2000), there were two independent radiations from Cuba to the Cayman Islands that included an earlier *lewisi* colonization of Grand Cayman and a more recent *caymanensis* colonization of Little Cayman and Cayman Brac. Interestingly, *lewisi* is as genetically distinct from *C. nubila* as it is from *C. cychlura*.

Conservation Status

The Grand Cayman Blue iguana is protected from international trade on Appendix I of CITES and is listed on the IUCN Red List as Critically Endangered (www.redlist.org). It is also listed as Endangered by the US Fish and Wildlife Service and is protected in the Cayman Islands by the Animals Law, section 68.

SISTER ISLES IGUANA (*Cyclura nubila caymanensis*)

Barbour and Noble, 1916 Synonyms: *Cyclura caymanensis*, Barbour and Noble, 1916; *Cyclura macleayi caymanensis*, Grant, 1940; *Cyclura nubila caymanensis*, Schwartz and Thomas, 1975

Description

The Sister Isles iguana (*C. n. caymanensis*) (Figure 2.27) is a large iguana that is characterized by four scales between the prefrontal shields and the frontal scale, five scale rows between the frontal and interparietal, and 10 scales between the anteriormost canthals. The first prefrontal is rarely in contact with the precanthal. It is characterized by eight vertical loreal rows, 11–14 superciliary scales, 12–16 subocular scales, and 64–82 dorsal crest scales (Schwartz and Carey, 1977). According to Burton (2004a), Schwartz and Carey identified no uniquely diagnostic scale characters. Burton notes that scalation of the head can sometimes distinguish *lewisi* from *C. nubila* and *C. n. caymanensis*; in most *caymanensis*, the fourth enlarged auricle spine is absent, and the second and third tend to be separated by one or more rows of small scales, leaving a gap between the third and fifth auricle. Also the second pair of prefrontals is absent in most individuals.

Burton (2004a) also states that the easiest way to distinguish *C. nubila*, *C. lewisi*, and *C. n. caymanensis* is by color. In *C. n. caymanensis*, adult males are typically light gray with tan on the head, tail, limbs, and dorsal midline of the body (Gerber, 2000a). The head and dorsal crest scales are often tinged in blue or reddish-pink hues. Diagonal black bars may ring the body and tail, but usually fade with age. The chest and belly may be burnt-orange or rust-colored. Females usually lack red or blue coloration and often have a greenish wash to the entire body. Like adult males, females have black front feet. Juveniles are tan

FIGURE 2.27 Adult female Sister Isles iguana (*Cyclura nubila caymanensis*).

or brown with 5–10 pale dorsal chevrons. These are bordered in black and break up laterally to form ocelli (Schwartz and Carey, 1977; Gerber, 2000a).

The generic name *Cyclura* is derived from the Greek words *cyclos*, meaning circular, and *urus*, meaning tail, after the thick-ringed tail characteristic of all iguanas in the genus. The specific name *nubila* is Latin for gray, but in this case refers to the name Gray rather than the color. John Edward Gray was the British zoologist who first described the Cuban iguana, *C. nubila*. The subspecific name *caymanensis* refers to the Cayman Islands, where the animal occurs.

Natural History Notes

The Sister Isles iguana is a very large iguana; adult males may reach 8.5 kg and 570 mm SVL or more. Females have been recorded up to 5.2 kg and 472 mm SVL (Gerber, 2000a). This iguana is known only from Little Cayman and Cayman Brac, which lie 100 km NE of Grand Cayman (Figure 2.28). These iguanas are found in coastal and interior habitats, but are most common in areas of limestone and dolomite covered in dry evergreen bushlands and thickets (Carey, 1966; Gerber, 2000a). Burrows and rock holes in these areas are used as retreats by iguanas.

Sister Isles iguanas feed primarily on plants and over 40 plant species have been identified in their diets, as well as land crab carcasses and slow-moving insects (Gerber, 2000a). Fruits and flowers are relished, and female iguanas can sometimes be found congregating in areas with fruiting bushes and trees. Native people on Cayman Brac have complained of iguanas raiding their gardens, eating especially the leaves of the potato plant (Carey, 1966) (Figure 2.29).

Very little research has been done on the Sister Isles iguana. Gerber (2000a) spent nine months in the field on Little Cayman and found that these iguanas compete intensely for

FIGURE 2.28 An aerial view of Little Cayman.

FIGURE 2.29 Mural on Little Cayman depicting the local iguana, with which islanders sometimes have to share their garden crops.

territories that are occupied during all seasons. The largest males occupy the best habitats and breed with numerous females. The smallest, youngest males do not hold territories and move throughout all of the territories, attempting to breed with any female while avoiding detection by dominant males. Females occupy smaller home ranges than males and can be territorial when nesting.

Males and females probably reach reproductive maturity in 2–3 years. Breeding females on Little Cayman measure 308–472 mm SVL (Gerber, 2000a). Sister Isles iguanas court and breed at the end of the dry season in April and May as temperatures and photoperiod increase. Females lay a single clutch of up to 25 eggs from late May to mid-June. Eggs are laid in a chamber 10–50 cm below the surface of the soil, often in small soil patches in the island interior. Alternatively, females may migrate considerable distances to the coast to lay their eggs in sandy soil. Females may guard their nests from conspecifics, especially other nesting females, for periods of from one day to several weeks. Hatchlings emerge from early August to early September after a mean incubation period of 72 days. Hatchlings average 107 mm SVL and 50.0 g in mass (Figure 2.30). Gerber (2000a) marked most juveniles he encountered in nests. Few were ever re-sighted, although one animal was later found 5.2 km from the nest site. Hatchling and juvenile Sister Isles iguanas grow 100 mm SVL per year for their first two years of life (Gerber, 2000a). The oldest known Sister Isles iguana was a captive specimen with a known age of 33 years (Slavens and Slavens, 1994).

Both Little Cayman and Cayman Brac have been inhabited by people since the early 1800s. In the 1930s and 1940s, iguanas were very common on both islands (Grant, 1940; Lewis, 1944), yet it was also reported that dogs were rapidly reducing both iguana populations (Grant, 1940). In 1965, Sister Isles iguanas were abundant only on a small section of coastal land on Cayman Brac (Carey, 1966), and today they are nearly extinct there. On Little Cayman the population appears stable, although in a patchy distribution of disjunct habitats (Gerber, 2000a).

FIGURE 2.30 Hatchling Sister Isles iguana.

Taxonomic Notes

The Sister Isles iguana is most closely related to the Cuban iguana (*C. nubila nubila*) and the Grand Cayman Blue iguana (*C. lewisi*), as well as the northern Bahamian iguanas (*C. cychlura*). According to Malone et al. (2000), there were two independent radiations from Cuba to the Cayman Islands that included an older *lewisi* colonization of Grand Cayman and a more recent *caymanensis* colonization of Little Cayman and Cayman Brac.

Conservation Status

The Sister Isles iguana is protected from international trade on Appendix I of CITES and is listed on the IUCN Red List as Critically Endangered (www.redlist.org). It is also listed as Threatened by the US Fish and Wildlife Service. Iguanas are protected in the Cayman Islands by the Animals Law of 1976.

CUBAN IGUANA (*Cyclura nubila nubila*)

Gray, 1831 Synonyms: *Cyclura nubila*, Cope, 1885; *Cyclura harlani*, Dumeril and Bibron, 1837; *Cyclura macleayi*, Gray, 1845; *Cyclura nubila nubila*, Schwartz and Thomas, 1975

Description

The Cuban iguana (*C. n. nubila*) (Figure 2.31) is a large species of iguana with three scales between the prefrontal shields and frontal scale, five scale rows between the frontal

FIGURE 2.31 Cuban iguana (*Cyclura nubila nubila*).

and interparietal scales, and six scales between the anteriormost canthals. The first prefrontal scale is rarely in contact with the precanthal scales. There are five vertical loreal rows, seven supralabials to the eye center, 9–15 superciliary scales, 10–16 subocular scales, and 52–85 dorsal crest scales (Schwartz and Carey, 1977). Burton (2004a) notes that scalation of the head can sometimes distinguish *C. n. nubila* from *C. lewisi* and *C. nubila caymanensis*, but that no head scale characters consistently distinguish them from one another.

Burton states that the easiest way to tell the three taxa apart is by coloration. Adult Cuban iguanas are dark gray to black dorsally, often dotted or stippled with tan. The head and tail are usually tan and the sides of the body often have vague diagonal remnants of the juvenile chevron pattern. The forefeet are usually black in color. Juveniles have strong chevrons on the body, are usually black and pale tan, with 5–10 pale chevrons. These chevrons often break up anteriorly into pale ocelli or oval fragments, all bordered in black. The tail is streaked in black and gray and may be vaguely ringed. The upper surface of the hind limbs is dark with scattered ocelli (Schwartz and Carey, 1977).

The generic name *Cyclura* is derived from the Greek words *cyclos*, meaning circular, and *urus*, meaning tail, after the thick-ringed tail characteristic of all iguanas in the genus. The specific name *nubila* is Latin for gray, but in this case refers to the name Gray rather than the color. John Edward Gray was the British zoologist who first described the Cuban iguana.

Natural History Notes

Cuban iguanas reach a very large adult size: males reach 510 mm SVL or more and can weigh over 9.0 kg in mass (A. Alberts, T. Grant, and J. Lemm, unpublished data). Females average 320 mm SVL and can weigh as much as 5.0 kg. Body size is sexually dimorphic.

The Cuban iguana is well-distributed throughout Cuba and is primarily found in dry coastal areas. Populations on the mainland have decreased dramatically in the past 20 years, although iguanas on numerous islets along the north and south coasts remain relatively safe (Perera, 2000). Iguana densities in Cuba vary with habitat and level of protection. Densities of 4.42 animals per hectare to 40 animals per hectare have been recorded (Perera, 2000). The population on the United States Naval Base at Guantanamo Bay was healthy and robust in the mid-1990s, with numbers estimated at 2,000–3,000 individuals at a density of 5.3 individuals per hectare (Alberts et al., 2001). Since that time, numbers have decreased significantly due to increased human activity (J. Lemm, personal observation). An introduced population of Cuban iguanas was established on Isla Magueyes in southwestern Puerto Rico in the mid-1960s (Rivero, 1978). This population has been growing for some time and was supposedly founded by a single pair of iguanas (Perez-Buitrago et al., 2006). In 2006, this population was estimated at 400–500 individuals, with a density of at least 62.5 iguanas per hectare (Perez-Buitrago et al., 2006).

Like other rock iguanas, the Cuban iguana is primarily herbivorous. Most plant parts are consumed. There have also been several reports of animal protein being consumed. The most commonly reported animal fed on by these iguanas is the crab *Cardisoma guandhumi* (Perera, 2000). Birds are also taken opportunistically (Gerber et al., 2002) and researchers at Guantanamo Bay have reported large Cuban iguanas climbing into mist nets to capture trapped birds. There has also been one report of cannibalism in which an adult female on Isla Magueyes chased, captured, and ate a hatchling iguana (Perez-Buitrago et al., 2006). In order to digest large quantities of plant material, it is necessary for iguanas to maintain high body temperatures. On Isla Magueyes, adult Cuban iguana body temperatures were recorded at a fairly constant 38.6 °C during the day (Christian, 1986a). Hatchling iguanas kept at different night-time temperatures (15, 25, and 35 °C), but at a constant daytime temperature of 35 °C, grew at different rates (Christian, 1986a). Hatchlings maintained at 35 °C ate more food, passed food through their system more rapidly and grew faster than hatchlings experiencing lower night-time temperatures. That Cuban iguanas often carry high external and internal parasite loads has been documented by a variety of Cuban biologists (Barus and Coy Otero, 1969; Cerny, 1969; Coy Otero and Hernandez, 1982; Barus et al., 1996).

Cuban iguanas often live in dense colonies. At Guantanamo Bay, researchers found that dominance was associated with body and head size, display behavior, testosterone levels, home range size, femoral pore size, and proximity to females (Alberts et al., 2002). In the breeding season, high-ranking males defended relatively small territories that overlapped the home ranges of one to four females (Figures 2.32, 2.33). Low-ranking males did not defend territories and moved extensively throughout large, loosely-defined territories. Non-ranking males resembled females in appearance and behavior, occupying peripheral home ranges to avoid aggression by larger, more dominant males. In the non-breeding season, home ranges of females and dominant males increased by 30%. During the subsequent breeding season, the five most dominant males were temporarily removed from the study site to see if lower-ranking males would adjust their home ranges and behavior, in turn increasing their reproductive success. Reproductive success was not measured directly, but the previously lower-ranking males assumed control of the best territories, won more fights, and gained access to more females.

FIGURE 2.32 Male combat in Cuban iguanas.

Cuban iguanas reach sexual maturity in 2–3 years. On Isla Magueyes, the smallest reproductive female weighed 983 g (Christian, 1986b). Mating takes place in May, with oviposition in June (Figure 2.34). A single clutch of up to 30 eggs is laid in soil or sand, often under objects such as rocks or plants. On Isla Magueyes, mean tunnel length is 153 cm with one or two turns. The average chamber depth is 43 cm. Only 37% of female iguanas guarded nests on Isla Magueyes, but it should be noted that some aspects of this population's behavior are

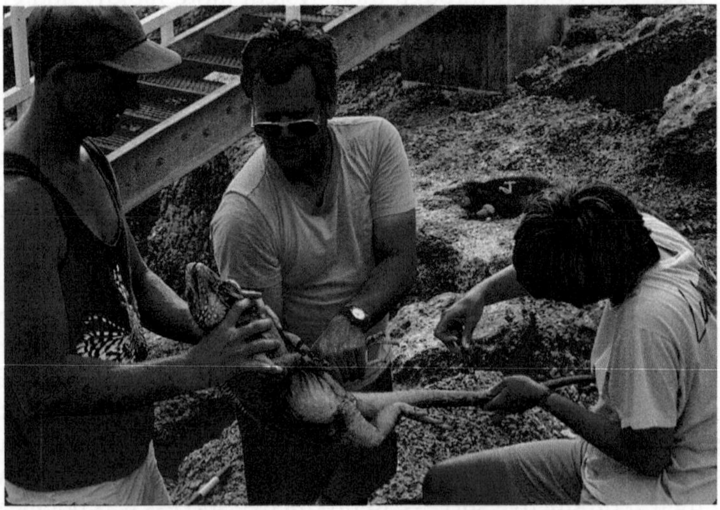

FIGURE 2.33 Jeff Lemm, Andy Phillips, and Allison Alberts take a blood sample from a wild Cuban iguana during one of their field studies of this taxon in Guantanamo Bay, Cuba, in the 1990s.

FIGURE 2.34 Wild adult Cuban iguana pairs are sometimes found together outside of the breeding season. This photo was taken in November, 2003.

known to differ from other sites in Cuba (Martins and Lamont, 1998). Wild nest temperatures have been recorded at 31–32 °C, with nests hatching in 70–78 days (Christian, 1986b). Eggs on Isla Magueyes average 43.5 mm long by 32.1 mm wide and weigh an average of 75.2 g; hatchlings average 9.9 cm SVL and an average mass of 48.2 grams (Christian, 1986b). Eggs from Cuban iguanas from Guantanamo Bay ranged in mass from 51.6 to 98.6 grams (Alberts et al., 1997). Egg lengths ranged from 64 to 82 mm and egg width ranged from 38 to 46 mm (A. Alberts and J. Lemm, unpublished data).

In an experimental study, Alberts et al. (1997) artificially incubated eggs from wild iguanas at Guantanamo Bay at three different temperatures (28.0, 29.5, and 31.0 °C) and at three different water potentials (−150, −550, and −1100 kPa). Water potential had no effect on hatchling growth or thermoregulatory behavior. Eggs incubated at 28.0, 29.5, and 31.0 °C hatched in 94, 108, and 128.5 days, respectively. Hatchlings from the various treatments ranged in mass from 43 to 55 grams and measured 9.4–10.2 cm SVL. Incubation temperature had little influence on size at hatching, but significantly influenced changes in body length, mass, and head dimensions, such that hatchlings incubated at higher temperatures grew faster in their first year. Studies have shown that hatchlings from drier nests have more fat and less yolk than hatchlings from wetter nests upon hatching (Christian et al., 1991). Growth rates of 8.37 mm SVL per month from 6 to 22 months have been calculated for Cuban iguanas (A. Alberts and J. Lemm, unpublished data).

Because the Cuban iguana is relatively stable in terms of population numbers, it is an ideal candidate for developing new conservation strategies. The Cuban iguana has been used successfully as a model for more critically endangered rock iguanas to test topics ranging from egg incubation techniques to the utility of headstarting (Alberts et al., 2004a).

Taxonomic Notes

The Cuban iguana is most closely related to the Sister Isles iguana (*C. nubila caymanensis*) and the Grand Cayman Blue iguana (*C. lewisi*), as well as the northern Bahamian iguanas (*C. cychlura*).

Conservation Status

The Cuban iguana is protected from international trade on Appendix I of CITES and is listed on the IUCN Red List as Vulnerable (www.redlist.org). It is also listed as Threatened by the US Fish and Wildlife Service.

ANEGADA ISLAND IGUANA (*Cyclura pinguis*)

Barbour, 1917 Synonyms: *Cyclura mattea*, Miller, 1918; *Cyclura portoricensis*, Barbour, 1919

Description

The Anegada Island iguana (*C. pinguis*) (Figure 2.35) is a large iguana that is characterized by relatively smooth skin and low, inconspicuous nuchal crest scales. It has 6–7 scales between the frontal and interparietal scales, azygous scales absent in the prefrontal suture, 10–12 scales between the anterior canthal scales, 7–9 scales bordering the frontal scale, and

FIGURE 2.35 Adult male Anegada Island iguana (*Cyclura pinguis*).

the first prefrontal never in contact with the precanthal. There are 3–10 vertical rows of loreal scales, 12–14 sublabials to the center of the eye, 1–6 postmentals, 54–63 dorsal crest scales, and nuchal crest scales never larger than body dorsal crest scales (Schwartz and Carey, 1977).

Adults are grayish to brownish-black dorsally with varying amounts of blue coloration on the dorsal spines, rear legs, and base of tail (Mitchell, 2000). Ventral coloration varies from white to light gray. Hatchlings and juveniles are gray or greenish with turquoise or gray vertical bars down the sides that are edged in black.

The generic name *Cyclura* is derived from the Greek words *cyclos*, meaning circular, and *urus*, meaning tail, after the thick-ringed tail characteristic of all iguanas in the genus. The specific name *pinguis* means fat, in reference to the animal's stocky appearance (often earning it the name Stout iguana).

Natural History Notes

The Anegada Island iguana is a large species; males can measure over 55 cm SVL and attain a mass of 8.0 kg (Carey, 1975), although more recent studies found that males average 45.0 cm SVL and 4.0 kg and females average 41.3 cm SVL and 2.9 kg (G. Gerber and J. Lemm, unpublished data). Fossils from caves on Puerto Rico and Amerindian kitchen middens from St Thomas show that the Anegada Island iguana was once distributed throughout the Puerto Rican Bank (Miller, 1918; Barbour, 1919; Pregill, 1981). Fossils in Puerto Rico and the Virgin Islands, once believed to represent the separate species *C. portoricensis* and *C. mattea*, were actually found to be *C. pinguis* (Pregill, 1981). Fossil remains from the late Pleistocene indicate that the Anegada Island iguana occurred on Puerto Rico 15,000–20,000 years ago when the islands of the Puerto Rican Bank were united as a single land mass due to lower sea levels associated with the last glacial maximum (Gerber, 2000b). In the cooler late Pleistocene, areas of Puerto Rico where iguana fossils have been found were believed to be much more xeric than they are today. A return to mesic conditions, combined with the fragmentation of Puerto Rico due to rising sea levels 6,000–8,000 years ago, probably resulted in the current restriction of this iguana to Anegada Island in the British Virgin Islands (Pregill, 1981). Until recently, the only surviving population was found on Anegada, a 4,000 hectare island on the northeastern extent of the iguana's historic range. Over the past 30 years, Anegada Island iguanas have been translocated to a number of islands in the British Virgin Islands. In the late 1980s, eight iguanas were moved from Anegada to Guana Island, where they thrived (Goodyear and Lazell, 1994). Since that time, progeny from these eight individual founders have been moved to at least five other islands (Mitchell, 2000; K. Bradley, personal communication).

On Anegada, iguanas typically utilize two habitat types: sandy scrub or rocky woodland. In sandy areas, iguanas use burrows that they dig themselves. In rocky areas, iguanas use self-dug burrows or natural holes in the limestone (J. Lemm, personal observation). The Anegada Island iguana is primarily herbivorous and feeds on a variety of species of plant leaves, fruits, and flowers. Germination studies have shown that seed germination is temporally enhanced by passage through the gastrointestinal tract of these iguanas (Gerber, 2000b).

Insects and crabs have also been found in their scat (Mitchell, 2000; G. Gerber and J. Lemm, unpublished data) (Figure 2.36).

Anegada Island iguanas occupy fairly large home ranges and males are quite territorial, especially during the breeding season. Male home ranges average 6.6 hectares and female home ranges average 4.2 hectares (Mitchell, 1999). Home ranges overlap and male activity centers are usually associated with home ranges of females. Sexual maturity probably occurs at 4–7 years of age, and the smallest wild female known to lay viable eggs measured 37.8 cm SVL and weighed 2.08 kg (G. Gerber, unpublished data). Mating occurs in May and June and females lay a single clutch of up to 20 eggs in June or July (Gerber, 2000b). Nesting occurs in sand, with tunnels up to 320 cm long and up to 90 cm deep. Nest chambers average 52 cm long by 29 cm wide by 14 cm high. Eggs are half-buried in loose sand with an air space above that is roughly half the height of the nest chamber (G. Gerber and J. Lemm, unpublished data). Females often guard the nest for up to two weeks. Eggs average 64.8 mm long by 45.0 mm wide and weigh an average of 62.7 g. Wild nest temperatures have been recorded at 30.4–31.0 °C, with eggs hatching in an average of 92 days (Figures 2.37 and 2.38). Hatchlings average 10.1 cm SVL and weigh an average of 47.0 g (Gerber, 2000b; G. Gerber and J. Lemm, unpublished data). On Guana Island, hatchlings average 10.8 cm SVL and 60 g in mass (Perry et al., 2007). Growth rates based on five field captures and 47 captive individuals were estimated to be approximately 50 mm SVL per year (Gerber, 2000b). As in all rock iguanas, thermoregulation is vital to the survival of the Anegada Island iguana and basking is a very common behavior. For 60 iguanas, cloacal temperatures ranged from 30.6 to 42.2 °C (87.1–108 °F). During optimal basking hours from 11 am to 3 pm, body temperatures averaged 39.0 °C (102.2 °F) (Gerber, 2000b).

FIGURE 2.36 Anegada Island iguanas often feed on crabs and insects, although this animal protein only makes up a small percentage of their diet.

FIGURE 2.37 These wild Anegada Island iguana eggs have hatched and have been filled with sand to show their size.

FIGURE 2.38 A wild Anegada Island iguana hatchling emerging from the nest chamber.

The Anegada Island iguana has been steadily declining since the 1960s and it is believed that the population on Anegada has decreased by 80%. Carey (1975) noted the lack of juvenile iguanas and estimated adult numbers to be 2.03 iguanas per hectare with an even sex ratio. However, two decades later Mitchell (1999) found adult iguana density to have decreased to 0.36 animals per hectare, with an adult sex ratio of two males to one female. Since the late 1990s, the scarcity of iguanas on Anegada has been consistently noted (K. Bradley, G. Gerber, and J. Lemm, unpublished data) and hatchling or juvenile iguanas are rarely, if ever, seen outside the September–October hatching season. The exception is one tiny 0.28 hectare cay in the middle of a salt pond. Hatchlings and juvenile iguanas are abundant on this cay throughout the year and it is believed that female iguanas swim there to nest. This cay is one of the only places that feral cats cannot reach, allowing young animals to remain safe there. Due to the extreme feral cat and ungulate problems facing Anegada, a headstart facility was constructed on the island in 1997.

Since 2003, Kelly Bradley from Fort Worth Zoo has been releasing headstarted Anegada Island iguanas from this facility and tracking their movements via radiotelemetry. Her project hopes to answer numerous questions, including finding the minimum body size at which juvenile iguanas can survive cat predation, as well as determining the optimal habitats for releasing juvenile iguanas (Bradley and Gerber, 2005). To date, over 120 iguanas have been returned to the wild (Figure 2.39) with a two-year average survivorship of about 85%. In 2009, a National Park proposal was approved that aims to protect the core iguana area and known iguana nesting areas on Anegada.

FIGURE 2.39 Rick Hudson of the Fort Worth Zoo releases a headstarted Anegada Island iguana to the wild.

Taxonomic Notes

The Anegada Island iguana is the most ancient of all the rock iguanas and is considered to be basal to the other species, as well as the most genetically unique (Malone and Davis, 2004). Genetic studies suggest that the ancestor of the Anegada Island iguana first dispersed from the Puerto Rican Bank to Hispaniola. From there it moved to islands north and west, diversifying into nine species and several subspecies.

Conservation Status

The Anegada Island iguana is protected from international trade on Appendix I of CITES and is listed on the IUCN Red List as Critically Endangered (www.redlist.org). It is also listed as Endangered by the US Fish and Wildlife Service.

RICORD'S IGUANA (*Cyclura ricordii*)

Dumeril and Bibron, 1837 Synonyms: *Aloponotus ricordii*, Dumeril and Bibron, 1837; *Hypsilophus ricordii*, Fitzinger, 1843; *Cyclura ricordii*, Cochran, 1924

Description

The Ricord's iguana (*C. ricordii*) (Figure 2.40) is a large iguana that has greatly enlarged, spinose scales at each caudal verticil (Ottenwalder, 2000b) and lacks a distinct row of scales between the infralabials and sublabials (Glor et al., 1998). Nasal scales are always in contact with each other, the rostral, and the postnasal. The frontal scale is separated

FIGURE 2.40 Ricord's iguana (*Cyclura ricordii*).

from the interparietal by 7–11 scale rows, the first prefrontal is never in contact with the precanthal, and the anterior canthals are separated by 9–17 scales. A row of subocular scales continues posteriorly to form a series of supratympanic scales and several rows of small scales occur between the suboculars and supralabials. There are 51–78 dorsal crest scales, with the crest scales on the neck usually longer than those on the body (Glor et al., 1998).

Adult Ricord's iguanas have a dorsal pattern that consists of 5–6 pale chevrons on a brown to gray background. Five of the chevrons continue onto the lighter ventral surface where they form bold, narrow lines. Chevrons, which are very distinct and bold in juveniles, are retained by adults.

The generic name *Cyclura* is derived from the Greek words *cyclos*, meaning circular, and *urus*, meaning tail, after the thick-ringed tail characteristic of all iguanas in the genus. The specific name *ricordii* refers to French biologist Alexandre Ricord, who first wrote of the species in 1826 and collected the holotype specimen.

Natural History Notes

The Ricord's iguana is a large species, with males reaching 510 mm SVL and females reaching 430 mm SVL (Ottenwalder, 2000b). They are found only on the island of Hispaniola in four isolated subpopulations. Three subpopulations are known from the Dominican Jaragua-Bahoruco-Enriquillo Biosphere Reserve, and a recently discovered subpopulation is known from Anse-a-Pitres, Haiti (Rupp et al., 2008). Throughout much of their range they are sympatric with the Rhinoceros iguana (*C. cornuta cornuta*) (Ottenwalder, 2000b). Ricord's iguana is strongly associated with thorn scrub woodlands, where optimal habitats contain low rainfall, alluvial powdery-clay soils, and low, forested hills (Ottenwalder, 2000b) (Figure 2.41). They prefer to dig their own burrows, which they continue to expand over time. Trees and rock cavities are also used as retreats. The diet of the Ricord's iguana consists of a wide variety of plants, including cactus (Figure 2.42). Insects and crustaceans are also eaten opportunistically (Ottenwalder, 2000b).

Home range size has not been studied in Ricord's iguanas. Sexual maturity is reached in 2–3 years. Mating takes place in early spring and females lay a single clutch of up to 20 eggs from March through mid-June. Most nesting coincides with the spring rainy season (Rupp et al., 2008). Nests are dug in fine, sandy soils and the egg chamber depth averages about 40 cm. Incubation temperatures have been recorded at 30-31°C and nests take 95–100 days to hatch (Ottenwalder, 2000b). Hatching occurs from late June to late September. In 2007, hatching success for 194 Ricord's iguana nests was 92.4% (Rupp et al., 2008). Hatchlings average 8.74 cm SVL and 30.0 g.

The current geographic range of the Ricord's iguana is estimated to be about 60% of its historic extent. Population declines beginning in the 1970s continue today as a result of habitat loss, competition and predation by feral and non-native animals, and hunting (Ottenwalder, 2000b). The Indianapolis and Toledo Zoos have been working with several Dominican conservation groups, primarily Grupo Jaragua and ZooDom, to implement the IUCN Iguana Specialist Group Species Recovery Plan for Ricord's iguana.

FIGURE 2.41 Habitat of Rhinoceros and Ricord's iguanas on Isla Cabritos, Dominican Republic.

FIGURE 2.42 Wild Ricord's iguana feeding on cactus fruit.

Taxonomic Notes

Ricord's iguana is most closely related to the Turks and Caicos iguana (*C. carinata carinata*) and less closely related to the Rhinoceros iguanas (*C. c. cornuta*). Although Ricord's iguanas are geographically closer to Rhinoceros iguanas than they are to Turks and Caicos iguanas, geographic barriers are believed to have existed on Hispaniola, resulting in the speciation patterns evident today (Malone et al., 2000).

Conservation Status

Ricord's iguana is protected from international trade on Appendix I of CITES and is listed on the IUCN Red List as Critically Endangered (www.redlist.org). It is also listed as Endangered by the US Fish and Wildlife Service.

WHITE CAY IGUANA (*Cyclura rileyi cristata*)

Schmidt, 1920 Synonyms: *Cyclura cristata*, Schmidt, 1920

Description

The White Cay iguana (*C. r. cristata*) (Figure 2.43) is one of three poorly-defined *C. rileyi* subspecies that are small in size compared to other rock iguanas. It has six rows of scales between the prefrontal shields and the frontal scale, 9–10 scales between the anterior canthals, and 6–9 scales bordering the frontal. It has poorly-defined verticil loreal rows (not countable), six lorilabial scales, nine superciliary scales, seven sublabials to the center of the eye, and three rows of scales between the infralabials and sublabials. It has a well-defined postsacral crest, consisting of 104–123 total dorsal crest scales, with 40–48 scales around the tail and five scales in the fifth caudal verticil (Schwartz and Carey, 1977). It also has 35–49 femoral pores, the fewest of the three subspecies (Carter and Hayes, 2004).

Adults are usually gray with brown to orange-brown vermiculations; the dorsal crest scales forelimbs and parts of the head typically also have some orange coloration. Juveniles lack dorsal chevrons or diagonal markings, but have pale grayish longitudinal markings.

FIGURE 2.43 White Cay iguana (*Cyclura rileyi cristata*). *Photo by Joe Wasilewski.*

The generic name *Cyclura* is derived from the Greek words *cyclos*, meaning circular, and *urus*, meaning tail, after the thick-ringed tail characteristic of all iguanas in the genus. The specific name *rileyi* refers to Joseph Harley Riley, an American ornithologist. The subspecific epithet *cristata* is Latin for crest, referring to the animal's significant dorsal crest spines.

Natural History Notes

The White Cay iguana is considered the smallest of the rock iguanas (Figure 2.44). Males reach 0.76 kg and 280 mm SVL, and average adult size for both sexes is 0.37 kg in mass and 201 mm SVL (Carter and Hayes, 2004). As in other rock iguanas, males are larger than females and have more prominent dorsal crests.

This subspecies is found only on a 15-hectare island known as White Cay (also called Sandy Cay), in the southern Exumas of the Bahamas. It is thought to have occupied many larger islands in the past and Amerindians apparently used it for food and funerary offerings (Hayes et al., 2004). White Cay iguanas prefer coastal rocky habitats (Figure 2.45) and are more abundant there than in the dense forests on the northwestern portion of the island (Hayes, 2000a). Rock holes and self-dug burrows are used as retreats. The White Cay iguana is primarily herbivorous and feeds on a variety of native plants.

Home range size for the White Cay iguana averaged 0.27 hectares for five male and two female iguanas (Hayes et al., 2004). Relatively large home ranges may be due to the lack of competition on the island. Females attain sexual maturity at 300 g and 200 mm SVL (Hayes et al., 2004). The mating season begins in May and runs through June. Oviposition usually

FIGURE 2.44 White Cay iguanas are considered the smallest of the rock iguanas. *Photo by Joe Wasilewski.*

FIGURE 2.45 Like other *Cyclura*, White Cay iguanas are often found in rocky limestone habitats near the coast. *Photo by Joe Wasilewski.*

occurs in July. Although very little is known of the reproductive ecology of the White Cay iguana, other members of the *rileyi* group lay small clutches with a maximum of six eggs. Other aspects of the ecology of the White Cay iguana are also likely to be similar to the San Salvador and Acklin's iguanas.

In the past 20 years, the White Cay iguana has declined drastically in numbers, primarily due to invasive species. Smuggling for the pet trade has also been an issue. Ongoing research on the White Cay iguana since 1993 has shown a heavily skewed sex ratio of 85% males (Hayes et al., 2004). In addition, a raccoon somehow arrived on the island and predated at least 14 adult iguanas before it was found dead. Conservation efforts include restoration of key habitat and nesting areas, as well as rat eradication.

Taxonomic Notes

The closest relatives to the *C. rileyi* group are the Grand Cayman Blue iguana (*C. lewisi*) and the Cuban and Sister Isles iguanas (*C. nubila*). Preliminary investigations have found no genetic variation among the three *rileyi* subspecies. It is possible that the populations only recently became separated, but more genetic work is needed to clarify the relationships among this group (Malone and Davis, 2004).

Conservation Status

The White Cay iguana is protected from international trade on Appendix I of CITES and is listed on the IUCN Red List as Critically Endangered (www.redlist.org). It is also listed as Threatened by the US Fish and Wildlife Service. In addition, all Bahamian rock iguanas are protected under the Bahamas Wild Animals Protection Act of 1968.

ACKLIN'S IGUANA (*Cyclura rileyi nuchalis*)

Barbour and Noble, 1916 Synonyms: *Cyclura nuchalis*, Barbour and Noble, 1916

Description

The Acklin's iguana (*C. r. nuchalis*) (Figure 2.46) is one of three poorly-defined *C. rileyi* subspecies that are small in size compared to other rock iguanas. The Acklin's iguana can be distinguished from the other *rileyi* by several scale features, including four scale rows between the frontal and prefrontal scales, three rows of loreal scales, and eight superciliary scales. Caudal verticals are not as enlarged as they are in either the White Cay or San Salvador iguanas, and the enlarged postsacral scales form a shorter row. The Acklin's iguana also has more femoral pores than the other two subspecies (41–55) (Schwartz and Carey, 1977; Hayes and Montanucci, 2000; Carter and Hayes, 2004).

Adults are dark black to bluish-gray in color with orange-brown to blackish marbling (Figure 2.47). The ventral surface is gray with orange-brown marbling on the chest. Juvenile Acklin's iguanas are very similar in coloration to San Salvador iguanas, but with a darker mid-dorsal zone in the dorsal pale band. Body color ranges from tan to black and the head is brownish. The venter is gray and the gular area is dirty yellow (Schwartz and Carey, 1977).

The generic name *Cyclura* is derived from the Greek words *cyclos*, meaning circular, and *urus*, meaning tail, after the thick-ringed tail characteristic of all iguanas in the genus. The specific name *rileyi* refers to Joseph Harley Riley, an American ornithologist. The subspecific epithet *nuchalis* means neck, and refers to the thick scalation around the neck.

FIGURE 2.46 Acklin's iguana (*Cyclura rileyi nuchalis*). *Photo by Joe Wasilewski.*

FIGURE 2.47 Adult female Acklin's iguana. *Photo by Joe Wasilewski.*

Natural History Notes

The Acklin's iguana (sometimes called the Watling Island iguana) is small in size; males reach only 36.0 cm SVL and 1.65 kg in mass. Body size and appearance are sexually dimorphic. The Acklin's iguana occurs naturally on Fish and North Cays, between Acklin's and Crooked Islands in the Bahamas. They were once found on Long Cay, Crooked Island, and Acklin's Island, but have since become extinct on these islands. A population exists on a small island in the Exumas Land and Sea Park that was introduced in the 1970s by a private individual who moved five adult animals. The remaining range of the Acklin's iguana represents only about 0.2% of its historic range (Hayes et al., 2004).

Acklin's iguanas live in a variety of habitats, ranging from low-lying scrub to limestone rocks and taller forested areas. Retreat burrows are dug in sand or rocky areas, but interestingly these iguanas prefer to spend the night under plants or leaf litter (Thornton, 2000). This subspecies, like all rock iguanas, is primarily herbivorous. Aggregate feeding behavior has been observed by Thornton (2000). A large male defended a single bush, while up to 12 females freely climbed into it to feed. Other males undetected by the dominant male were able to feed with the females among the branches.

Male Acklin's iguanas are territorial throughout the year, but become increasingly aggressive toward other males during the breeding season (Figure 2.48). Thornton (2000) tracked 10 female iguanas using radiotelemetry and found that gravid females ($n = 5$) utilized home ranges of 2,047 m^2 whereas non-gravid females ($n = 5$) utilized smaller home ranges that measured 397 m^2 (based on 23–37 fixes). About 30% of gravid iguanas traveled up to 1,000 m from their usual home ranges to oviposit eggs. Males have not been radiotracked, but probably have larger home ranges than females. Acklin's iguanas likely reach sexual maturity in 5–7 years. The smallest reproductive female captured weighed 260 g and measured 19.5 cm SVL (Hayes et al., 2004). Breeding

FIGURE 2.48 This very old, colorful, male Acklin's iguana shows the typical battle scars seen in older male rock iguanas. *Photo by Joe Wasilewski.*

takes place in late May and June. Females lay a single clutch of 2–5 eggs in July in nest burrows measuring up to 235 cm long and 22 cm deep. Females often spend the night in the nest burrow after laying, and will defend the nest for a few days following oviposition. Wild nest incubation temperatures vary from 25 to 33 °C, with the warmest temperatures occurring at night. Eggs measure an average of 55.4 mm long by 30.2 mm wide and weigh and average of 27.1 g (Thornton, 2000). Eggs generally hatch in late September and early October. Hatchlings are similar in size to those of the San Salvador iguana, which measure 8.2–8.5 cm SVL and weigh 21.5 g on average (Hayes et al., 2004).

Density and biomass of the Acklin's iguana are high and reflect the quantity and quality of food available where these iguanas exist (Hayes et al., 2004). Although these animals live in remote locations, they occur only on low elevation islands, making them vulnerable to hurricanes. Non-native rodents have also made their way to these islands, although their impact on iguana populations is not yet known.

Taxonomic Notes

The closest relatives to the *C. rileyi* group are the Grand Cayman Blue iguana (*C. lewisi*) and the Cuban and Sister Isles iguanas (*C. nubila*). Preliminary investigations have found no genetic variation among the three *rileyi* subspecies. It is possible that the populations only recently became separated, but more genetic work is needed to clarify the relationships among this group (Malone and Davis, 2004).

Conservation Status

The Acklin's iguana is protected from international trade on Appendix I of CITES and is listed on the IUCN Red List as Endangered (www.redlist.org). It is also listed as Threatened by the US Fish and Wildlife Service. In addition, all Bahamian rock iguanas are protected under the Bahamas Wild Animals Protection Act of 1968.

SAN SALVADOR IGUANA (*Cyclura rileyi rileyi*)

Stejneger, 1903 Synonyms: none

Description

The San Salvador iguana (*C. r. rileyi*) (Figure 2.49) is the largest of three poorly-defined *C. rileyi* subspecies that are small in size compared to other rock iguanas. This subspecies is defined by Schwartz and Carey (1977) as having five rows of scales between the prefrontal shields and frontal scales, seven scales between the anterior canthals, 5–7 scales bordering the frontal, four vertical loreal rows, five lorilabial rows, six supralabials to eye center, 10 superciliary scales, six sublabials to eye center, and two rows of scales between the infralabials and sublabials. It also has poorly defined postsacral crest scales and 35–51 femoral pores (Hayes, 2000b; Carter and Hayes, 2004).

The dorsal color of adult animals varies tremendously (Figure 2.50), ranging from red, yellow, or green to brown, usually punctuated with darker marking and fine vermiculations. Males are generally more vividly colored than females. Juveniles are brown or gray, often with a slightly paler mid-dorsal band having longitudinal stripes near the mid-dorsal crest. Juveniles lack bright colors and vermiculations (Hayes, 2000b).

The generic name *Cyclura* is derived from the Greek words *cyclos,* meaning circular, and *urus,* meaning tail, after the thick-ringed tail characteristic of all iguanas in the

FIGURE 2.49 San Salvador iguana (*Cyclura rileyi rileyi*). *Photo by John Binns IRCF.org.*

FIGURE 2.50 San Salvador iguanas such as this adult female have the most variable color pattern of all the *Cyclura. Photo by Joe Wasilewski.*

genus. Both the specific and subspecific names refer to Joseph Harley Riley, an American ornithologist.

Natural History Notes

The San Salvador iguana is a small iguana, with males measuring up to 39.5 cm SVL and mass averaging 0.70 kg for both sexes (Hayes et al., 2004). Males attain a larger body size than females and body size varies by island. The largest animals are from Low Cay, where average body mass is 1.5 kg (Hayes et al., 2004). The San Salvador iguana was once found throughout the main island of San Salvador, but is now largely restricted to two cays, Guana and Pigeon, within San Salvador's hypersaline lakes and four tiny offshore cays (Goulding, Green, Low, and Manhead). They are rarely encountered on the main island. Three additional populations have been recently extirpated from Barn, High, and Gaulin Cays.

The San Salvador iguana occupies a wide array of habitat types, including mangroves, low scrub, and tall forests, but they are locally most common in the vicinity of limestone rock outcrops and stands of sea grape (Hayes, 2000b). Retreats are self-dug in soil or under rocks. Like other rock iguanas, San Salvador iguanas are primarily herbivorous and feed on a wide variety of plants, including red mangrove and *Opuntia* cactus. Other food items include insects, land crabs, and birds. An adult female iguana was also observed killing and eating what was suspected to be one of her own hatchlings (Hayes et al., 2004).

Like all iguanas, San Salvador iguanas use well-defined home ranges and are territorial for at least part of the year. On Green Cay, male home ranges measured 0.43 hectares and female home ranges measured 0.62 hectares (Hayes et al., 2004). Similar to the White Cay and Acklin's iguanas, San Salvador iguanas probably reach sexual maturity in 5–7 years. On Green Cay, the smallest reproductive females captured weighed 300 g and measured

20.0 cm SVL. Males breed with multiple females in May and June, often through forced copulations, and defend females from other males. Females generally lay a single clutch of 3–10 eggs in July. Eggs are laid in egg chambers 18–28 cm below the soil surface at the end of tunnels that measure 30–116 cm in length. Females may defend nests depending on density of other nesting females (Cyril et al., 2001). Eggs measure an average of 53.4 mm by 29.5 mm and weigh an average of 27.7 g. Incubation lasts about 90 days. An emergent hatchling was captured that weighed 21.5 g.

Introduced predators and loss of habitat have significantly impacted the population of San Salvador iguanas. In addition, a number of hurricanes in the 1990s severely degraded habitat and nesting sites, and, although apparently no longer an issue, disease decreased some subpopulations. Assuming the main island of San Salvador no longer sustains a viable population, this iguana occupies only 0.2% of its former range (Hayes et al., 2004).

Conservation measures to date include eradicating introduced predators and restoring habitat on many cays, as well as evaluating iguana numbers around San Salvador and its cays (Hayes et al., 2004) (Figure 2.51). In 2005, 10 iguanas were moved from Green Cay to Cut Cay off San Salvador Island to create a new subpopulation.

Taxonomic Notes

The closest relatives to the *C. rileyi* group are the Grand Cayman Blue iguana (*C. lewisi*) and the Cuban and Sister Isles iguanas (*C. nubila*). Preliminary investigations have found no genetic variation among the three *rileyi* subspecies. It is possible that the populations only recently became separated, but more genetic work is needed to clarify the relationships among this group (Malone and Davis, 2004).

FIGURE 2.51 This adult male San Salvador iguana has had a bead tag sewn into his crest by researchers so that he can be identified at a distance. *Photo by Joe Burgess.*

Conservation Status

The San Salvador iguana is protected from international trade on Appendix I of CITES and is listed on the IUCN Red List as Endangered (www.redlist.org). It is also listed as threatened by the US Fish and Wildlife Service. In addition, all Bahamian rock iguanas are protected under the Bahamas Wild Animals Protection Act of 1968.

NAVASSA ISLAND IGUANA (*Cyclura cornuta onchiopsis*) – EXTINCT

Cope, 1885 Synonyms: *Cyclura onchiopsis*, Cope, 1885; *Cyclura nigerrima*, Cope, 1885; *Cyclura cornuta onchiopsis*, Schwartz and Thomas, 1975; *Cyclura onchiopsis*, Powell, 1999

Description

The Navassa Island iguana (*C. c. onchiopsis*) (Figure 2.52) is distinguished from the other two *C. cornuta* subspecies in that it had 30–44 dorsolateral scales in a distance equal to that between the nares and the eye (Powell, 2000b). It had two rows of scales between the prefrontal shields and the frontal scale, 4–6 scale rows between the supraorbital semicircles and the interparietal, seven supralabials to eye center, 6–10 sublabials to eye center, and 33–38 femoral pores (Schwartz and Carey, 1977). Body color varied from gray to green or brown.

The generic name *Cyclura* is derived from the Greek words *cyclos*, meaning circular, and *urus*, meaning tail, after the thick-ringed tail characteristic of all iguanas in the genus. The species name *cornuta* is from the Latin *cornutus*, meaning horned, and refers to the horned

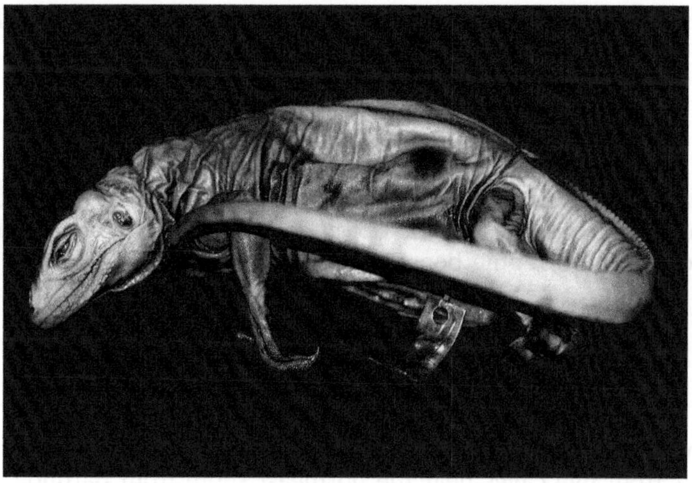

FIGURE 2.52 Specimen of the extinct Navassa Island iguana (*Cyclura cornuta onchiopsis*). *Photo by Bob Powell.*

projections on the snout. It is difficult to verify the origin of the subspecific name *onchiopsis*; it does not appear in Cope's first mention of this species (1885).

Natural History Notes

The extinct Navassa Island iguana was a large species. Maximum known size of adult females was 378 mm SVL, and maximum size of adult males was 420 mm SVL (Schwartz and Carey, 1977). This species was found only on Navassa Island, a small island off the west coast of Haiti. Nothing is known about the natural history of this iguana. It has been thought to be extinct for well over 50 years. Introduced feral animals, hunting, guano mining, and military occupation have all been blamed for the Navassa Island iguana's demise (Powell, 1999).

Taxonomic Notes

Cope (1885) first described this species as *C. nigerrima*, due to the animal's almost black coloration, and renamed it a year later as *C. onchiopsis*. Schwartz and Thomas (1975) reclassified it as a subspecies of *C. cornuta*. Many biologists, including Powell (2000b), believe that it should be elevated to species level based on morphology.

Conservation Notes

Cyclura cornuta onchiopsis is extinct. Because Navassa Island once supported a population of rock iguanas, there has been some discussion among conservation biologists about the potential for restoration of Navassa as a reintroduction site for one of the more critically endangered rock iguanas.

ADDITIONAL NOTES FROM THE FOSSIL RECORD

POSSIBLE UNNAMED SPECIES

In 1937, Allen wrote of remains of a mammal found in a cave deposit on Little Exuma Island in the Bahamas. Found with these *Geocapromys* bones were bones from "a few birds, reptiles, and frogs." These reptile bones were never described and were apparently lost (Pregill, 1982).

UNNAMED SPECIES

From 1958 to 1963, field teams from the University of Florida collected large numbers of vertebrate fossils from five caves near the northeast shore from Barbuda, Lesser Antilles. The late Pleistocene iguanid bones they found cannot be classified as *Cyclura* or *Iguana*, and most closely resemble the genus *Conolophus* from the Galapagos.

UNKNOWN SPECIES

Fossils from a large Pleistocene *Cyclura* were found in a cave known as Banana Hole, New Providence, Bahamas, in the late 1950s. Both Etheridge (1966) and Pregill (1982) examined samples from this cave. Etheridge examined body vertebrae that were believed to come from iguanas estimated to be 345–460 mm SVL. Pregill examined a fossil dentary bone from an iguana estimated to be 425–450 mm SVL. He suggested that the bones could have possibly been from an extinct population of *C. cychlura*.

CHAPTER

3

Natural History

CHAPTER

3

Natural History

OUTLINE

Habitat Requirements and Home Range	77	Social Behavior	87
Diet and Foraging	80	Reproduction and Life History	89
Predators and Defense	84		

HABITAT REQUIREMENTS AND HOME RANGE

West Indian iguanas occupy subtropical habitats from the Bahamas to the southern Greater Antilles. These islands are characterized by vegetation ranging from moderately dry forest to much drier cactus habitats. They are non-volcanic in origin, made up of heavily eroded limestone. Iguanas are most often found in coastal lowlands, generally comprised of limestone formed during times of high sea levels. Iguanas are heavily dependent on these limestone formations for retreats. In fact, one of the defining characters of rock iguanas, the comb scales on their feet, are evolved for climbing rocks (much like crampons used by mountaineers) (Figure 3.1). The red sclera of the eyes of most rock iguanas may have evolved for living on the bright white limestone that reflects the sun's rays, in order to help protect the eyes from solar damage (Figure 3.2).

Vegetation is another major limiting factor for the survival of iguanas. Without forage plants, iguana populations cannot survive. In turn, the tropical dry forests on these islands would not exist as they do today without iguanas to serve as seed dispersers. Studies have shown that seeds that have passed through the intestinal tract of iguanas germinate faster than seeds that have not been consumed (Hartley et al., 2000), and that seeds left in iguana scat produce seedlings that grow twice as quickly as seeds removed from iguana scat (Alberts, 2004). A final limiting factor for the viability of iguana populations is suitable nesting habitat, given that most West Indian iguanas require relatively deep soil in which to nest (Wiewandt, 1977; Iverson, 1979).

Cyclura, *First Edition* DOI: 10.1016/B978-1-4377-3516-1.10003-2

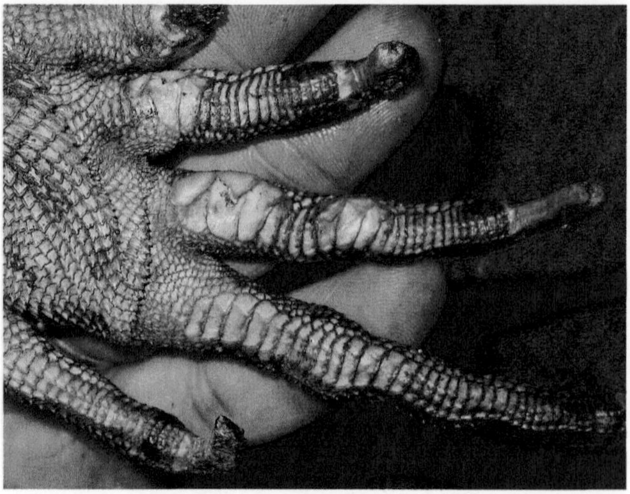

FIGURE 3.1 The comb scales on the rear toes of rock iguanas evolved for rock climbing.

FIGURE 3.2 The red sclera of the eyes of rock iguanas may serve to protect the eyes of rock iguanas from solar radiation.

Home range size and ecology vary tremendously among rock iguanas. These lizards are found on islands as small as 10,000 m^2 in size to areas over 110,000 km^2 (e.g., Cuba). On small islands, iguana home ranges are constrained by island size and iguana density (Knapp and Owens, 2005), whereas on larger islands, iguana home ranges are probably regulated by the distribution of conspecifics and key food resources.

FIGURE 3.3 Male Cuban iguanas fight over a female on a territorial boundary during the breeding season.

Because male rock iguanas are often territorial and polygynous (Perry and Garland, 2002), males typically have larger home ranges than females, although female San Salvador Island iguanas were found to have larger home ranges than males in a short study by Hayes et al. (2004). It is generally believed that males expand their home ranges in the breeding season as they search for mates. Some of the variation in home range size among species may be the result of logistical issues such as the limitations of radiotransmitter attachment methods or monitoring periods that are too short to adequately assess an animal's entire range. Male home range sizes have been found to range from 0.3 hectares in the Allen Cays iguana (Knapp, 2000b) to 11.53 hectares in the Andros iguana (Knapp, 2005a).

Daily movements made by iguanas are generally dictated by season and food availability. Outside the breeding season, iguanas usually remain in a core area or loosely defined territory. Thermoregulation and feeding usually occur in the morning and later afternoon, while midday is spent in the shade or in burrows. In some species, iguanas will move away from their burrows to access specific foraging areas. Cuban iguanas will travel as far as 500 meters from their coastal rocks and burrows to feed in adjacent thorn forest, and Turks and Caicos iguanas often travel as far as 300 meters from their burrows to feed on the shoreline or in nearby forests (J. Lemm, personal observation).

During the breeding season, male iguanas move extensively in search of mates, and females often travel great distances to nest. Both of these activities may result in overlapping territories and aggressive encounters between conspecifics (Figure 3.3). In other species, there is significant territorial overlap without aggression. This commonly occurs when iguanas exist at high densities, and usually in areas where artificial food supplementation by people occurs. Large numbers of Cuban iguanas and Allen Cays iguanas congregate in groups at sites where people feed them (Lacy and Martins, 2003; Hines, 2007).

DIET AND FORAGING

Rates of energy utilization by reptiles are generally quite low compared to mammals and birds of equal size (Christian et al., 1986), although on some islands with high lizard population densities, the rate of respiratory metabolism is quite high (Bennett and Gorman, 1979). Christian et al. (1986) estimated a high rate of energy expenditure in Cuban iguanas on Isla Magueyes. Adult males had daily expenditures averaging 245 kJ, and adult females and juveniles had daily energy expenditures averaging 171.5 kJ and 111.6 kJ, respectively. The large average energy expenditure of the Cuban iguana population on Isla Magueyes – approximately 4800 kJ/(ha × d) – is believed to be the result of large body size and high population density. These estimates are greater than those for Galapagos land iguanas, small lizards, and small mammals (Christian et al., 1986), and are possibly the result of the high body temperature preferences of Cuban iguanas, their activity rates, and/or their large body size (Nagy, 1982; Snell and Christian, 1985). Given their large body size and high preferred body temperature, rock iguanas need large quantities of food to forage, interact, avoid predators, and reproduce.

Rock iguanas have relatively high metabolic rates and need fairly large volumes of food to satisfy their energetic expenditures, despite the fact that they spend quite a bit of time resting. Auffenberg (1982) showed that Turks and Caicos iguanas spend only 18% of their day feeding and/or foraging, yet it only takes a short time to fill the digestive tract with vegetation and the proximal colon always contains digesta (Iverson, 1982). Many iguanas weigh less in the summer than they do in the spring. In winter, much of their time is devoted to foraging (Iverson, 1979; Auffenberg, 1982). Reduced summer feeding is probably related to reproductive behavior, a pattern that has also been noted in captive iguanas (J. Lemm, personal observation).

All large iguanas are adapted for a life of herbivory, and lizard herbivory probably had its roots in xeric habitats. To gain enough energy to live on plants, lizards had to adapt to solve problems arising from salt imbalances and seasonal difficulties in obtaining adequate animal food (Iverson, 1982). Most herbivorous lizards – and all rock iguanas – have extrarenal osmoregulatory organs, or salt glands (Hazard, 2004). These nasal salt glands allow herbivorous and omnivorous lizards that feed on potassium-rich plants to excrete excess sodium. Rock iguanas have from five to nine valves in the proximal colon to handle the increased bulk of vegetative material that comes with an herbivorous lifestyle (Iverson, 1982) (Figure 3.4).

The larger the species, the more complex the colon. This specialized colon slows digestion time, which increases absorption of nutrients and houses the high nematode loads that aid in digestion (Iverson, 1982). Iverson (1979) estimated the number of nematodes in a single healthy adult Turks and Caicos iguana to be 15,000. Juveniles begin accumulating colic nematodes soon after hatching, and worm populations usually number at least 100 by the time the animal is three months old. Worms are ingested when an animal tongue flicks the ground. Many young iguanas are known to eat the feces of adult iguanas, possibly to purposely ingest these worms to aid in digestion. Iverson (1982) also found that the number of nematodes increases with the number of colic valves, and suggests that colon complexity has allowed iguanas to reach very large adult body sizes, which in turn reduces predation probability and provides metabolic and thermoregulatory benefits.

FIGURE 3.4 A male Andros iguana forages in the late afternoon.

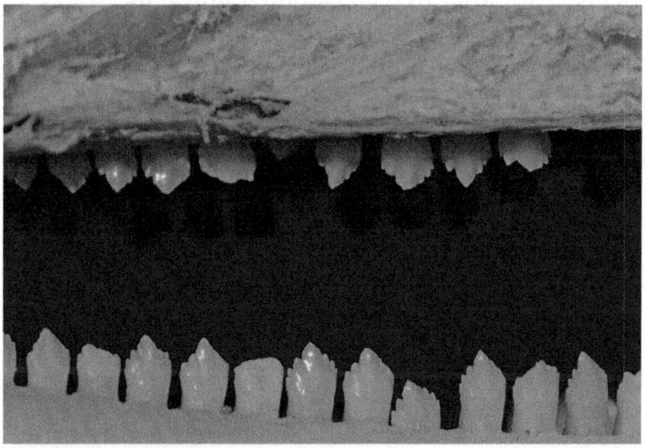

FIGURE 3.5 Teeth of an adult Cuban iguana.

Rock iguanas have teeth that have evolved for shearing vegetation (Figure 3.5). They have pleurodont, laterally-compressed, multicusped teeth and their jaw structure precludes mastication of food (Bjourndal, 1997).

This results in the ingestion of large food particles, and often whole leaves passing through the intestinal tract intact (Iverson, 1982). The cell walls of plants contain fibrous materials that are normally difficult to digest, but which contain structural carbohydrates that can be digested by some microbes (Troyer, 1983). In the absence of oxygen, these microbes, usually bacteria and sometimes protozoans, ferment structural carbohydrates and release volatile fatty acids as waste products. These are readily absorbed through the gut tissues and are used for energy and synthetic processes (Troyer, 1983). By housing microbes, iguanas gain energy and nutrients that would normally be unavailable. McBee and McBee (1982) found that through hindgut fermentation, the green iguana (*Iguana iguana*) acquired as much as 40% of its energy needs, compared to 9–19% in hindgut-fermenting mammals.

Rock iguanas are primarily herbivorous and eat the stems, leaves, buds, and flowers of a variety of plants, as well as fungi such as mushrooms (Auffenberg, 1982). Most species have been shown to feed on some animal matter as well (Wiewandt, 1977; Iverson, 1979; Auffenberg, 1982; Goodman, 2007). Small insects are often consumed accidentally, but larger insects and other invertebrates, lizards (including juvenile rock iguanas), crabs, fish, and birds all fall prey to iguanas (Figures 3.6–3.8). At Guantanamo Bay, Cuba, a large male Cuban iguana was observed climbing into mist nets set by bird researchers and was consuming every bird caught in the nets.

Rock iguanas have also been recorded eating pieces of their own shed skin, and feeding on the feces of conspecifics or other animals (Iverson, 1979; Coenen, 1995; Knapp, 2001; Goodman, 2007). The reported number of plant species consumed varies greatly by species and by island. Hayes et al. (2004) found that San Salvador iguanas on some islands eat only seven types of plants, whereas Grand Cayman iguanas eat as many as 105 different species (Burton, 2010). Iguanas have also been known to feed on one species of plant for a few days, and then switch to another species (Iverson, 1979; Auffenberg, 1982). This behavior is thought to maximize food diversity in order to gain essential nutrients.

Food diversity is quite important to rock iguanas. Auffenberg (1982) found that Turks and Caicos iguanas on Pine Cay utilized 62–91.3% of the plants occurring there, but that a greater proportion of plants were eaten in simple habitats compared to diverse ones. Hayes et al. (2004) found that male San Salvador iguanas weighed more on islands with greater food diversity and taller vegetation, although this could be the result of lower population density. In the Turks and Caicos and the Bahamas, reintroduced populations of rock iguanas grow much more rapidly than source populations and reach greater sizes, probably due to the fact that they have little to no competition (Knapp, 2001; Gerber, 2007).

FIGURE 3.6 Insects are commonly found in the feces of *Cyclura* – the feces pictured here are from an Anegada Island iguana (*C. pinguis*) and show the remains of many large bees.

FIGURE 3.7 Turks and Caicos iguana cannibalizing a hatchling. *Photo by Joe Burgess.*

FIGURE 3.8 Cuban iguana feeding on the carcass of a road-killed hutia (*Capromys pilorides*).

PREDATORS AND DEFENSE

Rock iguanas are often the largest vertebrate species on the islands where they occur, and therefore as adults rarely face predation from native animals. In contrast, hatchling and juvenile iguanas are commonly eaten by birds and snakes, and occasionally even fish. Ospreys (*Pandion haliaetus*) (Hayes et al., 2004) and other predatory birds, including merlins (*Falco columbarius*), peregrine falcons (*Falco peregrinus*), kestrels (*Falco sparverius*), and red-tailed hawks (*Buteo jamaicensis*), are known to feed on rock iguanas. A variety of shorebirds, including gulls (*Laridae* spp.) and herons and egrets (*Ardaidea* spp.), probably consume young rock iguanas as well.

Snakes are well-known predators of rock iguanas. Predation by racers (*Alsophis* spp.) (Knapp et al., 2010; J. Lemm, personal observation) and boas (*Epicrates* spp.) is common. Knapp and Owens (2004) were able to track six different Bahamian boas after they consumed juvenile Andros iguanas fitted with radiotransmitters. Cuban boas reach large sizes (up to 4.85 m) and have been known to feed on adult female Cuban iguanas (J. Lemm, personal observation).

Fish probably do not take a heavy toll on rock iguana populations, but Hayes et al. (2004) mention that a barracuda ate a young Acklin's iguana as it swam across a channel in the Bahamas. Land crabs (*Cardiosoma* and *Geocarcoidea* spp.) are common in the Caribbean and reach very large sizes. They have been known to feed on rock iguana eggs (J. Lemm, personal observation), although it is not known if the eggs were fertile or if the crabs ever capture and eat hatchlings. Finally, adult iguanas have been known to feed on hatchlings of their own species. Such cannibalistic encounters have occurred in Turks and Caicos (Iverson, 1979; Auffenberg, 1982) and San Salvador iguanas (Hayes et al., 2004).

People have predated on rock iguanas for centuries. There is abundant evidence that native inhabitants of the Caribbean, including the Arawak Indians, commonly hunted rock iguanas for food (Woodley, 1980). Rock iguana bones have been found throughout the

FIGURE 3.9 Feral cats and Indian mongoose (*Herpestes auropunctatus*) are both voracious predators of iguanas.

Caribbean in ancient kitchen middens. Today, rock iguanas are illegally hunted in a few locations, including Andros Island (Knapp, 2005b).

Introduced species are perhaps the greatest challenge to the survival of rock iguanas (Figure 3.9). Predators such as cats and dogs prey heavily on iguanas and have been known to cause local extinctions of iguanas. In the 1970s, a population of 15,000 Turks and Caicos iguanas was almost completely destroyed within five years by a few dogs and cats brought to Pine Cay by hotel workers (Iverson, 1978, 1979). In addition, the range of the Turks and Caicos iguana has been reduced to less than 5% of its original extent largely due to the introduction of predators (Welch et al., 2004). For much of the last century, the Jamaican iguana was thought to be extinct as a result of the introduction of the Indian mongoose (*Herpestes auropunctatus*) to Jamaica. Today, the mongoose is considered to be the most important factor preventing recruitment of juveniles into the breeding population (Wilson et al., 2004a).

Other non-native animals that negatively impact on rock iguana populations include European pigs (*Sus scrofa*), rats (*Rattus rattus* and *R. norvegicus*), raccoons (*Procyon lotor*), and mice (*Mus musculus*). Pigs are thought to eat rock iguana eggs and possibly hatchlings, and to destroy nesting habitat. Wiewandt (1977) reported a 25–100% annual loss of Mona Island iguana eggs to feral pigs. In 1997, a raccoon reached White Cay and killed at least 14 adult iguanas, bringing mortality of the White Cay iguana that year to 35–67% (Hayes et al., 2004). Mice and rats have been discussed as possible predators of rock iguana eggs and hatchlings, although the evidence for this is circumstantial. Auffenberg (1982) believes the extirpation of San Salvador iguanas on High Cay may have been due to rat infestation. Introduced rodents are found on many of the islands where rock iguanas occur.

FIGURE 3.10 Goats and other feral hoof stock eat the same plants as iguanas and trample nesting habitat. These goats were photographed in the core iguana area of Anegada.

Herbivorous non-native mammals have also become a problem in the Caribbean (Figure 3.10). Hoof stock such as donkeys, cows, goats, and sheep are often allowed to freely roam islands and denude vegetation at alarming rates. The resultant competition for food with iguanas is believed to put some rock iguana populations at risk (Carey, 1975; Mitchell, 1999). Introduced hoof stock also trample iguana nesting areas and burrows (G. Gerber, personal communication; J. Lemm, personal observation).

Rock iguanas have few defenses against introduced predators. Simply taking cover in rocks, burrows, or trees is often not enough to evade dogs, cats, and mongooses. When entrenched in burrows and rock crevices, iguanas often fill their bodies with air or use their tails to wedge into small spaces and can be killed by mammals attempting to extricate them from their refuge sites. Iguanas generally exhibit a flight-or-fight response when faced with predators. Even juvenile iguanas will stand high on their feet, fill their bodies with air, turn sideways to look larger, hiss, and charge predators when cornered.

Hatchling iguanas often take to the trees and hide for their first months of life, but still succumb to birds and snakes. Adults of the larger species of rock iguanas, especially males, have little to fear other than dogs and large snakes. When a large Cuban boa approached a colony of adult Cuban iguanas at Guantanamo Bay, several large male iguanas approached the snake, bobbing their heads, gaping, hissing, and nipping at the snakes tail, effectively driving it off (J. Lemm, personal observation) (Figure 3.11). Young iguanas use a different strategy, filling themselves with air to try to prevent smaller snakes from swallowing them, but will succumb to larger snakes (Figure 3.12). Perhaps the best way young iguanas can keep themselves safe from predators is by choosing the proper habitat. Knapp et al. (2010) observed that newly hatched Andros iguanas spent more time in open mangrove

FIGURE 3.11 This adult Cuban iguana effectively drove off this large Cuban boa (*Epicrates angulifer*) through a series of displays and aggressive actions.

FIGURE 3.12 Hatchling telemetered Andros iguana being consumed by a Bahamian racer (*Cubophis (= Alsophis) vudii*). *Photo by Chuck Knapp.*

habitat, where survival was higher and predators were less common, than in closed-canopy forest.

SOCIAL BEHAVIOR

Rock iguanas are social animals that live in either large group settings (e.g., Bahamian iguanas, Turks and Caicos iguanas) or in temporary aggregations with more dispersed lifestyles (e.g., Jamaican iguanas, Mona Island iguanas). In either social setting, there must be a communication system to decrease aggression, yet still maintain a healthy, breeding population.

Rock iguanas communicate in a number of ways, both chemically and visually. The femoral pores on the underside of the hind legs secrete a mixture of lipids and proteins that are thought to be used to attract mates and mark territories (Alberts, 1993). Iguanas use their tongues and the Jacobson's organ in the roof of the mouth to identify scent marks. This process, called chemoreception, delivers chemical signals from the tongue directly to the brain via the vomeronasal nerve. Iguanas may also use feces to communicate. It is quite common to observe iguanas tongue-flicking the feces of conspecifics, and chemical signals in feces potentially communicate territory, reproductive condition, and health and body condition of the signaler.

The other primary mode of communication in rock iguanas is visual signals. The most important form of communication is the head bob, in which the head is raised and lowered in a stereotypical, species-specific bobbing pattern (Martins and Lamont, 1998). These displays are used in almost all social interactions and vary according to the behavioral context in which they are used. Head bobs are most often seen in the aggressive encounters that accompany territorial maintenance, and during the breeding season. In general, males tend to display much more frequently than females (Martins and Lacy, 2004). In some species, appeasement head bobs are employed to reduce aggression in group settings (Iverson, 1979; Martins and Lacy, 2004). Displays vary between species, populations, and even individuals, often containing information about an animal's identity (Martins and Lamont, 1998).

Rock iguanas also use body posturing to communicate. In aggressive encounters and territorial disputes, large males will often turn their bodies sideways to appear larger to conspecifics. They may also stand high on their feet and fill their bodies with air to look as large as possible. Dorsal crest spines on the back not only help iguanas acquire more ultraviolet light to aid calcium absorption, but also make them appear larger during aggressive encounters. Body posturing is also observed in submissive displays, usually in females and low-ranking males. Such displays usually involve the iguana getting its body as close to the ground as possible, as if trying to hide. The base of the tail is often raised, possibly allowing dominant animals to tongue-flick the vent area in order to assess breeding condition or social status (J. Lemm, personal observation).

Fights are common, especially during the breeding season as males defend territories and females. Head bobs, body posturing, and gaping serve as a first defense, but often

FIGURE 3.13 Gaping is a common form of aggressive display in rock iguanas.

escalate into physical pushing matches to contest territories or mates. Fighting is often observed on the boundary between two territories during the breeding season. Male Cuban iguanas will sit face to face, gape, hiss, wriggle the tail tip, and lunge at one another (J. Lemm, personal observation) (Figure 3.13). Fights are usually short-lived and primarily involve tail whips to the body and jaw-wrestling, in which both males have their mouths open, grasping each other's jowls as they push each other back and forth. The thorn-like projections on the faces of male rock iguanas serve as protection in these instances. On very rare occasions during intense fights when neither iguana backs down, males bite one another with tremendous force, potentially breaking tails and sometimes severing limbs in very bloody battles. On one occasion, a male Cuban iguana even pushed a rival male off a cliff to his death (J. Lemm, personal observation).

Hatchling and juvenile iguanas show similar behavior to adult iguanas. Dominance is established early in life, with the larger animals usually emerging as the most dominant. Head bobbing, gaping, jaw wrestling, and fighting are commonly observed under both captive and wild conditions.

REPRODUCTION AND LIFE HISTORY

It is unclear at what age rock iguanas begin to establish territories, but young iguanas seem to prefer certain areas after they disperse. In both Cuba and the Turks and Caicos Islands, juvenile iguanas were observed on a daily basis to be using the same general home range (J. Lemm, personal observation). As rock iguanas grow and socialize, rank and social status are established. Because many species can reach sexual maturity by two years of age (Burton, 2010; G. Gerber, personal communication), setting up territories and finding mates is important.

Rock iguanas are polygynous, meaning that males typically breed with more than one female. The most dominant males breed with the most females. In Cuban iguanas, dominance is associated with body and head size, display behavior, testosterone levels, home range size, and proximity to females. An experimental study conducted by Alberts et al. (2002) showed that when the five highest-ranking males were removed from an area, previously lower-ranking males assumed control of vacated territories, won more fights, and increased their proximity to females.

In Cuban iguanas, males are either high-ranking territory holders, low-ranking peripheral/marginal males, or non-ranking males, also called sneaker males (Alberts et al., 2002). These sneaker males look and act like females, seeking to remain undetected by dominant males and even successfully reproduce on occasion. Dominant males establish territories within a home range during the breeding season in conjunction with chasing off rival males and fighting with other high-ranking males. Males generally attempt to mate with as many females as possible, whether through territorial defense (high-ranking males), occupying the periphery of defended territories and attempting forced copulations (low-ranking males), or sneaking copulations when dominant males are occupied elsewhere (non-ranking males). Females are not normally territorial and do not fight with one another.

Males court females with head bobbing and often tongue-flick the feces or vent of females to assess whether or not they are in breeding condition. Females who are ready to breed relay

FIGURE 3.14 A pair of Cuban iguanas copulating while another female tries to attract the attention of the male. The numbers on the sides of the animals were applied by researchers.

this message with head bobs and body posturing. Females who are not ready to breed will also actively be chased, and if caught, may be forced to copulate with males. Even non-forced copulation looks quite rough: the male grabs the female by the top of the neck with his mouth, and using his rear legs, attempts to lift the rear legs of the female in order to gain access to her cloaca (Figure 3.14). When the cloacas are aligned, the male everts one of two hemipenes and copulates with the female. Mating is rarely observed in pairs more than once or twice a day and copulation lasts anywhere from 30 seconds to 10 minutes or more (Blair, 1991; J. Lemm, personal observation).

After a female has bred, she generally remains in her territory and may breed with additional males. About a month after breeding, gravid females begin to look for areas to nest. A few days before nesting, gravid females often stop feeding. Rock iguana eggs are large and take up a great deal of space in a female's abdomen (Alberts, 1995), and as much as 50% of a gravid female's body mass can consist of eggs. Some rock iguana females dig their nests in their home ranges, while others make migrations to nest, some as far as one kilometer (Thornton, 2000), often to traditional communal nest sites where many females gather to nest.

Most rock iguanas prefer to nest in areas of open soil (Iverson et al., 2004), although some species lay in shallow nests in the sand (Hayes et al., 2004), and one species, the Andros iguana, has been documented to lay its eggs in termite mounds (Knapp et al., 2006). Females often dig several test holes before they nest. Some of the larger species lay eggs in chambers at the end of deep tunnels, which can be over a meter in length with many tunnels (Iverson et al., 2004). Female rock iguanas usually dig a separate burrow to oviposit, although Iverson (1979) found that Turks and Caicos iguanas often lay eggs in side tunnels of active retreat burrows. Female Cuban, Sister Isles, and San Salvador iguanas have been shown to close the entrance to the nest burrow behind them during the final stages of oviposition (Iverson et al., 2004).

Egg and clutch sizes vary across species. Turks and Caicos iguanas lay an average of five eggs that weigh an average of 26 g (G. Gerber, personal communication), whereas Mona Island iguanas average 11 eggs per clutch that average 104 g in weight (Wiewandt, 1977; Wiewandt and Garcia, 2000). In most species, clutch mass is strongly correlated with female body size and larger females lay earlier than smaller females (Iverson et al., 2004).

Some rock iguana species defend nests prior to oviposition, whereas almost all female rock iguanas defend nests after oviposition for at least a few days (Iverson et al., 2004). Female rock iguanas choose nesting sites with an average temperature of about 30 °C (86 °F). At this temperature (which fluctuates with time of day in the wild), eggs generally hatch in about 100 days. Survivorship of wild nests in seven taxa ranged from 76 to 85%, with excess soil moisture and predation being the main causes of nest failure (Wiewandt, 1977; Iverson et al., 2004).

Hatchling size varies across species. Turks and Caicos iguanas weigh about 15 g when they hatch, whereas Mona Island iguanas weigh about 75 g at hatching. After hatching, young rock iguanas sometimes stay in the nest to absorb any residual yolk and rest for several days to several weeks (Christian, 1986b), but must eventually reach the surface and disperse. Most hatchling rock iguanas emerge from the nest via a tunnel that leads directly to the surface. Once the surface is reached, iguanas disperse and often climb into trees for cover (Christian, 1986b; Perez-Buitrago and Sabat, 2007).

Dispersal is dangerous for young iguanas, and mortality may be quite high in the first few weeks of life. Dispersal has only been studied in three species of rock iguanas and survivorship differed greatly. Mona Island iguana hatchling survivorship was estimated to be only 22% (Perez-Buitrago and Sabat, 2007), whereas Iverson (2007) estimated 97% survival for the Allen Cays iguana. Young iguanas dispersed in a straight line away from nests in all cases; some moved as far as 5.08 m away from nests (Mona Island iguanas, Perez-Buitrago and Sabat, 2007), while others dispersed up to 600 m (Andros island iguanas, Knapp et al., 2010). Knapp et al. (2010) found that dispersal into mangrove habitat improved hatchling survival and hypothesized that areas with lower habitat heterogeneity may increase hatchling survivorship due to reduced predator abundance and diversity.

Many rock iguana hatchlings are naturally preyed upon by snakes and birds, which likely cue in on nesting areas during the hatching season (Knapp et al., 2010) (Figure 3.15). The unusually high survivorship for Allen Cays hatchlings may result from a lack of snake predators and the fact that hatchlings cannot disperse very far on small cays. As rock iguanas mature, survivorship increases. In the Turks and Caicos iguana, survivorship is roughly 55% for the first three years of life, 67% for years four through six, and 90–95% for adults (Gerber and Iverson, 2000).

For an iguana to grow and mature, it needs to avoid predation, yet still feed and thermoregulate. The larger an iguana gets, the fewer predators it has, so rapid growth is very important. Feeding and thermoregulation, the two main factors affecting growth, vary greatly. Growth rates across species are also related to body size, with smaller species growing faster (Iverson, 1989). For captive animals, growth rates vary tremendously based on the care they are given. Alberts et al. (1997) found that Cuban iguana eggs incubated at

FIGURE 3.15 Snakes such as this Anegada racer (*Cubophis* (=*Alsophis*) *potoricensis anegadae*) prey heavily on juvenile iguanas.

higher temperatures resulted in more rapidly growing hatchlings in the first year. However, after 16 months of age, growth rates no longer varied among hatchlings incubated at different temperatures.

Rock iguanas typically grow quickly in the first year and then growth slows with age. Iverson (1979) showed that Turks and Caicos iguanas grew 2 cm SVL/year in year one, whereas adults grew only 0.2–1.7 cm SVL/year (Gerber and Iverson, 2000). Wiewandt (1977) estimated growth rates of 5.3 cm SVL/year for first-year Mona Island iguana hatchlings. Iverson (1989) found that growth rates in the Allen Cays iguana were faster for males (1.764 cm SVL/year) than for females (1.139 cm SVL/year). Iverson and Mamula (1989) also showed that in the first year of life, irrespective of maximum body size, rock iguanas grew more slowly than mainland forms of iguanas, suggesting that island-dwelling iguanas may be more food-limited than mainland forms.

Interestingly, iguanas that have been translocated to new islands grow more rapidly than the source population from which they came. Knapp (2001) found that repatriated Allen Cays iguanas and their offspring grew faster than iguanas on the source island. Translocated Turks and Caicos iguanas also grow more rapidly than iguanas from the source populations. Gerber (2007) found that offspring born on translocation cays grew up to four times faster and sexual maturation decreased from 6–7 years to 1.5–2.5 years. It is believed that lack of competition and increased food availability are responsible for increased growth and maturation rates among translocated iguanas.

Iguanas tend to grow very slowly once they reach adulthood, although they continue to grow throughout life. The oldest known iguana on record was a male Grand Cayman Blue iguana named Godzilla. He was 69 years of age when he died at the Gladys Porter Zoo in Texas. The oldest known female iguanas are still being studied by John Iverson. These wild Allen Cays iguanas have been known to reproduce at 39.8 years and possibly even up to 61.5 years (in 2002) (Iverson et al., 2004). It is believed that rock iguanas can live 60–80 years, although reproductive senescence likely occurs earlier than this. Alberts et al. (1997) found that largest, oldest-appearing female Cuban iguanas produced a greater number of infertile eggs and exhibited a higher egg mortality rate than did smaller, younger females.

CHAPTER

4

Husbandry

CHAPTER

4

Husbandry*

OUTLINE

Population Management	97	Capture, Restraint, and Handling	109
Quarantine	98	Reproduction and Nesting	112
Housing	98	Hatchling Care	117
Feeding	104	Record Keeping	123

POPULATION MANAGEMENT

Since 1995, a studbook has been maintained for all rock iguana species kept in American Zoo and Aquarium (AZA) institutions. The captive history of rock iguanas held in more than 50 institutions from 1898 to the present is recorded in the studbook, as well as information from some private breeders. The primary goal of the studbook is to better mange rock iguana breeding efforts in North America and to keep track of the genetic relatedness of captive animals. Animals in headstarting and breeding facilities in Anegada, Jamaica, Grand Cayman, and the Dominican Republic are also included in the studbook.

The studbook software SPARKS (Small Population Animal Record Keeping System) is used to document key aspects of population management, ranging from growth records and medical data to facility transfers. Additional software (PM2000) is used in conjunction with studbook data to formulate a carrying capacity for the captive population and provide management recommendations.

*The majority of the information presented in this chapter is based on the authors' experiences with both wild and captive rock iguanas over the past 20 years. These husbandry protocols have been refined at the San Diego Zoo's Kenneth C. and Anne D. Griffin Reptile Conservation Center, where we maintain and breed four species of rock iguanas.

Cyclura, *First Edition* DOI: 10.1016/B978-1-4377-3516-1.10004-4

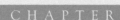

FIGURE 4.1 Digger, a Grand Cayman Blue iguana from the Blue Iguana Conservation Program breeding facility.

AZA also develops and implements Species Survival Plans (SSPs) for endangered species. The primary goal of an SSP is to manage genetically diverse stock in captivity as a hedge against extinction in the wild. Participating institutions play a role in generating support for conservation and research, as well as species recovery in the wild. Since the Rock Iguana SSP was approved in 1996, over 20 US institutions have become involved and are contributing funds to the program. In addition, many of these facilities are directly involved in fieldwork and recovery efforts in the wild. Currently, the Rock Iguana SSP is focused on three critically endangered species, the Grand Cayman Blue iguana, the Anegada Island iguana, and the Jamaican iguana (Figure 4.1).

QUARANTINE

In order to reduce the spread of parasites and other ailments in captive facilities, animals new to a facility should be quarantined for 30–90 days in separate holding areas. Physical examinations, standard blood tests, and at least three fecal samples testing negative for endoparasites are typically required at most facilities prior to a new animal joining other collection animals. Quarantine is also the best time to implant a passive integrated transponder (PIT) tag for identification purposes if the new animal does not yet have one.

HOUSING

Because most West Indian iguanas occupy large areas in the wild, spacious enclosures are necessary for their captive care. Pairs should be kept in enclosures at least 10 ft × 6 ft × 6 ft (3 m × 1.8 m × 1.8 m), but larger areas (minimum of 12 ft × 12 ft × 10 ft or 3.6 m × 3.6 m × 3.0 m) are ideal. Adult males should never be housed together because they will almost always fight, leading to severe injuries.

Climbing is important for iguanas of all ages, and cage height should be adjusted so that keepers can comfortably carry out their work. Cage heights of 8–10 foot (2.4–3.0 m) are easy to access and allow lizards to maintain a comfortable level of security. Climbing structures such as large logs and shelves are also useful for both juvenile and adult iguanas. Hide areas are also necessary for successfully maintaining rock iguanas, especially if animals are not able to burrow. Retreats, visual barriers, and hides such as PVC tubes, concrete blocks, large rocks, plants, wood shelters, logs, ice chests, and a variety of other retreats have been used by many institutions with success. Burrows provide the best shelter and allow rock iguanas to dig as they would naturally. If given the proper substrate depth, rock iguanas will dig very deep, long burrows. These burrows give animals a place to spend the night, seek cover, and thermoregulate. A range of rock iguana enclosures are shown in Figures 4.2–4.7.

Plants provide excellent cover and browse. Native West Indian plants such as sea grape provide cover and are durable because iguanas do not normally feed on their leaves. Readily available nursery plants such as *Ficus* and *Hibiscus* can withstand enclosure temperatures and provide cover, but adult iguanas will denude the leaves so quickly that plants may need to be rotated in and out of enclosures every week or two. The most ideal, readily available plants in the United States that provide sufficient cover and sight barriers are species such as palms that are rarely browsed in captivity.

Enclosures should be designed so animals can be separated if needed. Two smaller enclosures separated by an access door have proven useful. In some facilities, smaller animals may become stressed near larger animals or may be constantly harassed by them. Smaller animals should be moved into cages with smaller conspecifics or housed alone. Animals that are thin,

FIGURE 4.2 Griffin Reptile Conservation Center, San Diego Zoo Safari Park.

FIGURE 4.3 Jamaican iguana display enclosure. Hope Zoo, Kingston, Jamaica.

show small injuries and scrapes on the neck, side, or feet, or constantly hide in group situations, are candidates for cage moves. Injured or sick animals should be housed alone and should always be monitored when being reintroduced to their cage mates, as resident iguanas may not welcome them back.

Most enclosures are constructed of pressure-treated lumber frames or galvanized steel frames covered with half-inch by one-inch mesh. The mesh size can be finer for smaller animals, but larger animals often catch toes or tear nails in enclosures made of smaller meshes. Ideally, the largest mesh that will prevent animal escape and intrusion by snake predators, rats, and mice should be used. Mesh-covered walls should have a strip of plastic or wood at the base to keep animals from rubbing their noses or fighting with neighboring animals. Smooth concrete walls (at least 4 feet high on the inside) are ideal for rock iguana enclosures, and concrete, glass, and/or plastic materials (PVC sheets) can be used for indoor housing. For outdoor areas, a combination of low wall and mesh is inexpensive and practical.

Digging barriers should be placed under the enclosure substrate to prevent escape via burrowing. An ideal substrate depth is 3–4 feet of soil or sandy soil that will maintain humidity and burrow integrity. Some institutions use pea gravel or decomposed granite, and this works well; however, animals can construct more suitable burrows in dirt. In addition, some facilities report intestinal impactions from pea gravel, decomposed granite, and sand. Humidity levels of 50–80% are ideal for keeping most iguanas and if necessary can be maintained throughout the year by lightly spraying enclosures with water and keeping

FIGURE 4.4 Anegada Island iguana exhibit, San Diego Zoo.

FIGURE 4.5 Indoor/outdoor iguana enclosures at the Fort Worth Zoo's Animal Outreach and Conservation Center. *Photo by Rick Hudson.*

FIGURE 4.6 Cuban iguana outdoor enclosure, San Diego Zoo Institute for Conservation Research (animals from these older enclosures were later moved to the Griffin Reptile Conservation Center).

FIGURE 4.7 Cover, in the form of hides and burrows, is particularly important to the well-being of all rock iguanas. This is a Jamaican iguana.

plants and planters moist. Water should be provided at all times and cages should be cleaned daily, being careful to remove all feces, especially near feeding areas.

Ultraviolet light is necessary for processing vitamin D3 and promoting proper bone mineralization. Indoor/outdoor enclosures are useful in this regard, allowing animals to receive natural, unfiltered light for at least part of the year. In colder regions, UV-transmitting plastics and glass can be used for skylights. Some facilities are using Solacryl SUVT™ panels (Polycast Technology Corp., Stamford, CT) that have approximately 85% UVB transmission. Recently, however, it was discovered that Solacryl may be similar to UVB bulbs in that animals have to be in close proximity to the material to access usable UVB. A UV-transmitting plastic called Acrylite OP-4 (CYRO Industries, Rockaway, NJ) is being used by some facilities with success. Active UV Heat, once known as the Westron Dragonlite (T-Rex products, Inc. Chula Vista, CA), has proven to raise circulating D3 levels in other lizard species when suspended 200 cm above animals (Gillespie et al., 2000). Many facilities that house rock iguanas are now using these or other types of mercury vapor lamps and/or fluorescent bulbs (Figure 4.8). Unfortunately, some of these bulbs degrade very quickly and may even lack UVB right out of the box. A new bulb on the market, made by MegaRayUV (www.reptileuv.com), is said to provide very large amounts of UVB light and is purported to last for more than a year. Basking lights should be placed above basking platforms such as rocks, cinder blocks, logs, or shelves, as rock iguanas seem to feel more comfortable basking in elevated areas.

Rock iguanas require thermal gradients and high basking temperatures (Figure 4.9). Ambient enclosure temperatures should not drop below 65 °F (18.3 °C) at night and should not rise above 90 °F (32.2 °C) during the day. Optimal ambient temperatures should be maintained at roughly 85 °F (29.4 °C), with cooler areas for the animals to retreat. During winter, ambient temperatures can safely be lowered by about 5–7 °F (2.8–3.9 °C). Basking areas with

FIGURE 4.8 Rock iguanas require ultraviolet light, either by artificial lighting or natural sun.

FIGURE 4.9 Rock iguanas prefer to bask at high temperatures. This wild Turks and Caicos iguana is having his body surface temperature read in the field with a non-contact thermometer.

high temperatures should be maintained throughout the year. Wild animals typically maintain body temperatures of 96.8–105.8 °F (36–41 °C) throughout the day (G. Gerber, unpublished data). Captive specimens often bask at surface temperatures as high as 150 °F (65.5 °C).

It is interesting to note that the facilities with the most successful breeding programs utilize high temperature basking. These high temperatures can be safely provided using spotlight or floodlight-type heaters, as well as infrared brooders. Injuries such as thermal burns have never been recorded when following recommended heating protocols. High-temperature contact heaters, such as pig blankets, hot rocks, and heating pads, are not normally recommended for diurnal basking lizards, but some facilities use them without incident. It is believed that most burns occur when ambient conditions are cool and a cold animal is allowed to bask by sitting directly on a hot heat source, or when an animal is allowed to approach a basking source too closely, such that it is focused on a small part of the body rather than the entire animal.

FEEDING

Animals should ideally be fed on a daily basis, with fresh water available in water dishes at all times. Dishes should be used for food so that substrates, which may be responsible for gut impactions, do not mix with the food. One dish should be provided per animal, even when housed in pairs. Paired animals will sometimes fight over food, so it may be necessary to provide a visual barrier between dishes when animals are feeding (Figure 4.10). Hatchling animals may share a large dish. If a larger, dominant animal is observed chasing subordinates from the feeding area, the smaller animal should be moved or provided with another dish.

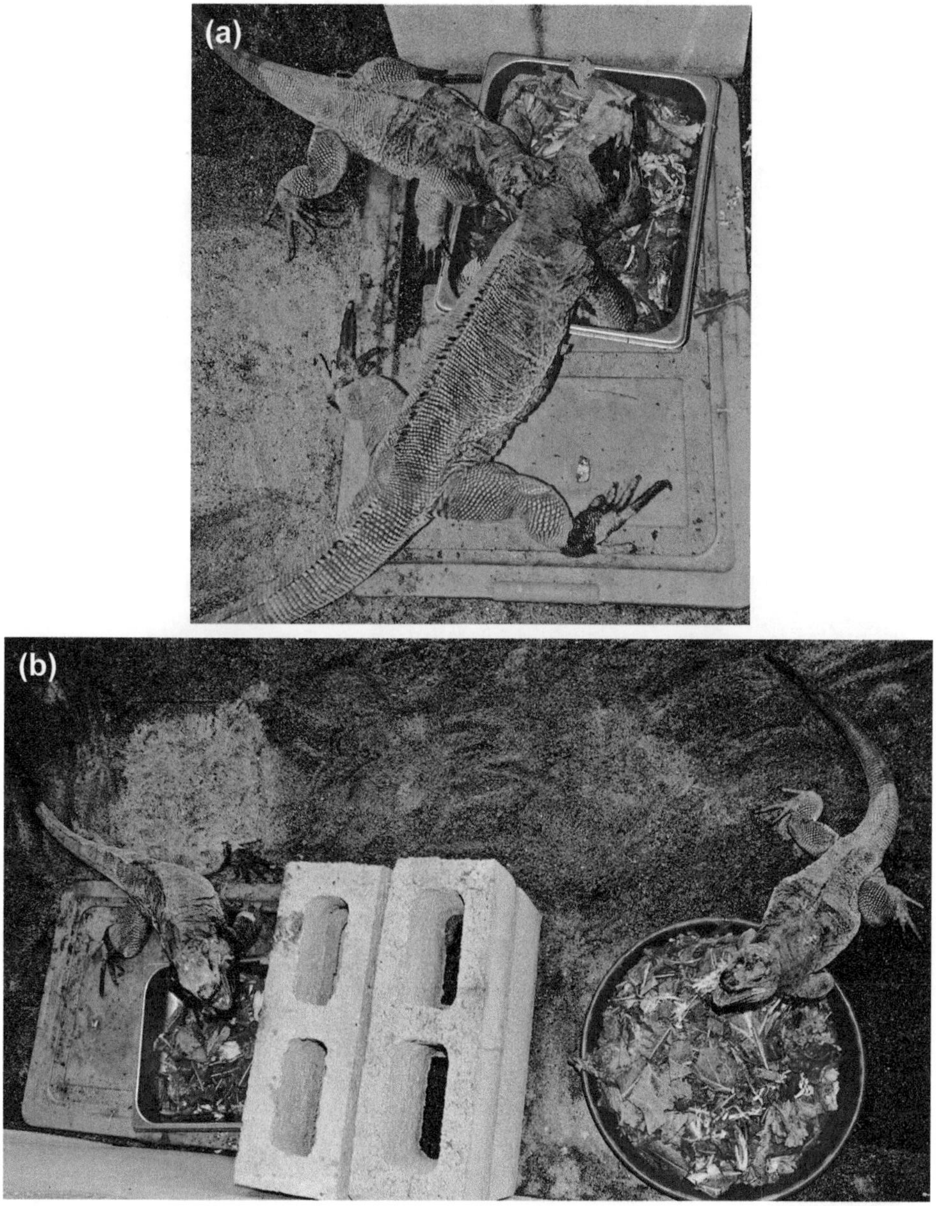

FIGURE 4.10 Conspecific aggression often occurs during feeding, as seen in this pair of Anegada Island iguanas (*a*). Plates can be separated with a visual barrier to reduce this aggression (*b*).

FIGURE 4.11 Ingredients for iguana diets at the Griffin Reptile Conservation Center. These ingredients should be mixed thoroughly so iguanas do not pick out favored foods so easily.

FIGURE 4.12 Rock iguanas require a great deal of food. In the wild they often browse through the cooler parts of the day.

Greens are chopped according to the size of the animals being fed. Adults are fed chopped greens measuring about 3 × 3 inches (7.6 × 7.6 cm), whereas hatchlings are fed finely chopped greens measuring roughly ¼ × ¼ of an inch (0.63 × 0.63 cm). Juveniles are fed greens chopped to a size of 1 × 1 inch (2.54 × 2.54 cm). Fruits and vegetables should be finely grated and mixed in with the other dietary items, with seeds and cores removed before grating. It is important to mix the food well so that the animals cannot easily pick out the favored foods (Figures 4.11, 4.12).

Recommended diets for adult and subadult rock iguanas consist of a variety of vegetables, fruits, and supplements (Tables 4.1, 4.2). At the San Diego Zoo's Griffin Reptile Conservation Center, daily greens are supplemented with a 50:50 mix of high fiber herbivore pellets and Leaf eater food (Marion Zoological, Plymouth, MN). The high fiber mix is included not only as a supplement, but also to avoid loose stools. The fiber is mixed with water, but if fecal deposits are still loose, the amount of water can be decreased until the stools are firmer. Sometimes it is necessary to eliminate the water in the mix, and in these cases, the grated fruits and vegetables are mixed directly in with the fiber diet, then added to the greens and mixed. In addition, a gel mix is offered that provides a small amount of animal protein (Table 4.1). On occasion, insects such as crickets can be fed to hatchling rock iguanas. Some institutions offer pinky or fuzzy mice to their animals on rare occasions. Although all taxa of rock iguanas are known to feed on some animal matter, wild rock iguanas feed predominately on vegetation and

TABLE 4.1 Standard Daily Diet for One Adult Iguana at the San Diego Zoo's Griffin Reptile Conservation Center

Item	Weight	Day
Collard greens	125 g	Mon, Wed, Fri
Mustard greens	125 g	Mon, Wed, Fri
Chard	125 g	Mon, Wed, Fri
Dandelion greens	125 g	Mon, Tue, Wed, Thu, Fri
Kale	125 g	Tue, Thu
Bok choy	125 g	Tue, Thu
Escarole	125 g	Tue, Thu
Root vegetable (various)	15 g	Mon, Tue, Wed, Thu, Fri
Squash (various)	15 g	Mon, Wed, Fri
Green beans (chopped)	15 g	Tue, Thu
Fruit (various)	15 g	Mon, Tue, Wed, Thu, Fri
High fiber diet (ground)[1]	35 g	Mon, Tue, Wed, Thu, Fri
Reptile Carnivore/Omnivore Gel[2]		

[1]*Mixed 1 : 1 with water by weight.*
[2]*Components: Turtle Brittle (Nasco International, Fort Atkinson, WI); Leafeater Diet (Marion Zoological, Plymouth, MN); gelatin (dry, unsweetened); carrots (raw); greens (raw; kale, collard, dandelion, mustard). Preparation (1 kg gel): 200 g Nasco Turtle Brittle (ground), 45 g Knox gelatin, 90 g chopped leafy greens, 90 g chopped/grated carrot, 575 g hot water. Adapted from information provided by the Tennessee Aquarium.*

very little is understood about animal proteins in their diet. Too much animal protein is believed to cause serious health problems.

On some days of the week, especially in the spring and summer, browse plants, usually *Hibiscus* and *Ficus*, as well as mulberry (*Morus*), are offered to the iguanas (Figure 4.13). These are simply freshly cut branches that are placed in PVC "vases" that hold water in the bottom. The "vase" is bolted to the walls of the enclosure and the branches are placed inside. All iguanas, including hatchlings, will feed on the leaves, flowers, and fruits of the plants. Some edible potted plants are kept in the enclosures and when the leaves are sufficiently browsed clean, the plant is replaced with a fresh one.

A key component of the standard daily diet described here for an adult iguana is Reptile Carnivore/Omnivore Gel (see Table 4.1, note 2). Preparation of this gel is as follows:

1. Add prepared greens and carrots to a high-power blender, followed by all the dry ingredients.
2. Add hot water to the blender. Immediately homogenize all the ingredients for 3 minutes. The mixture should be a thick liquid.
3. Pour the mixture into a shallow pan and allow the gel to set in a refrigerator.
4. Cut the gel with a knife or food processor to obtain appropriately sized pieces.
5. Keep the gel refrigerated and use within seven days. The gel can be frozen in an airtight container for up to three months.

TABLE 4.2 Standard Daily Diet for One Subadult Iguana at the San Diego Zoo's Griffin Reptile Conservation Center

Item	Weight	Day
Collard greens	60.0 g	Mon, Wed, Fri
Mustard greens	60.0 g	Mon, Wed, Fri
Chard	60.0 g	Mon, Wed, Fri
Dandelion greens	60.0 g	Mon, Tue, Wed, Thu, Fri
Kale	60.0 g	Tue, Thu
Bok choy	60.0 g	Tue, Thu
Escarole	60.0 g	Tue, Thu
Root vegetable (various)	7.5 g	Mon, Tue, Wed, Thu, Fri
Squash (various)	7.5 g	Mon, Wed, Fri
Green beans (chopped)	7.5 g	Tue, Thu
Fruit (various)	7.5 g	Mon, Tue, Wed, Thu, Fri
Miner-All O (without D3)[1]	1/8 teaspoon	Mon, Wed, Fri
Miner-All I (with D3)	1/8 teaspoon	On the 1st and 15th of each month

[1]*Sticky Tongue Farms/Miner-All, Sun City, CA.*

FIGURE 4.13 Anegada Island iguana browsing on *Hibiscus*.

Note that additional items can be added to the gel in preparation, including mealworms, extra chopped fruits and/or vegetables, and crushed limestone or oyster shell. The gel can also be used as a vehicle for medications.

The gel should be fed free choice as the primary diet. Consumption will vary by species. Uneaten gel should be removed on a daily basis.

CAPTURE, RESTRAINT, AND HANDLING

In order to reduce the risk of injury to both animals and keeper, proper restraint techniques are necessary whenever iguanas are handled. It should be noted that handling and restraint put a tremendous amount of stress on an animal and iguanas should only be handled when necessary (veterinary exams, weighing and measuring, moving to new enclosures). In addition, keepers should do everything possible to reduce stress on cage mates that are not being captured. Captive animals may show signs of stress for a few days immediately following capture, including appetite loss, constant hiding, and flight behavior.

Because iguanas have strong jaws and large, powerful claws, minor injuries to keepers are common. Even juvenile iguanas can inflict bite wounds that require stitches, and a bite from an adult iguana can be serious. Iguana scratches are common, especially from the long rear toes. In addition, hatchling and juvenile iguanas may have their tails broken off due to improper handling techniques.

Perhaps the easiest, least-stressful way to capture a non-tractable animal from within an enclosure is with a net. Large fishing nets, with the netting replaced by a sturdy cloth bag, work well for this purpose. Use of a cloth bag is important because netting rips easily, leading to escape, and iguanas may become tangled in the netting. "Hand-grabbing" or manually capturing iguanas works well with younger animals or larger, non-aggressive adults. When cornered, some individuals may become very agitated and rush or jump toward the keeper with open mouths.

In some cases, tamer individuals can be handled without restraint for educational purposes and related activities. The easiest way to handle these individuals is by resting them on the forearm with the hand gently supporting the chest of the animal (Figure 4.14). The handler's second hand is used to support the rear of the body. If the animal should become agitated, the forward hand can easily be shifted to restrain the head, while the other hand can restrain the tail and/or rear legs. Because these tame animals can quickly become nervous or agitated outside of their normal quarters, handlers should always be aware of their surroundings and potential escape hazards.

Hatchling iguanas should be restrained in the middle of the body with the head secure (Figure 4.15). Larger juvenile and adult iguanas should be restrained with two hands (Figure 4.16). One hand should lightly yet firmly grasp the animal behind the head, either in the neck or shoulder region to prevent the animal from turning and biting. The second hand should be placed over the pelvic region, keeping a firm grasp on the rear legs to

FIGURE 4.14 Allison Alberts demonstrates the correct way to handle a tractable iguana.

FIGURE 4.15 The proper handling of a juvenile iguana.

FIGURE 4.16 The two-person method of holding and restraining of a large rock iguana.

prevent scratches to the hand or arm that is restraining the head. The tail of the iguana is also a powerful weapon and can be restrained under the arm of the hand that is grasping the rear legs. In many cases it is easier to restrain the tail and one of the rear legs using the same hand. When possible, large iguanas should be restrained by two people, with one person holding the head and a second person holding the hindquarters of the animal.

During measurements of larger iguanas, some keepers and field researchers have found that blindfolds such as elastic knee bands placed over the entire head of the animal work well to calm the animal and keep it from attempting to bite. Because many species will keep the mouth agape during restraint and any hand movement near the mouth may result in a bite, this technique will also help protect the person taking measurements. In addition, many researchers simply turn an iguana over on its back, which generally quiets the animal significantly, even if it is highly agitated.

REPRODUCTION AND NESTING

With the exception of the Rhinoceros iguana and the Cuban iguana, successful reproduction of rock iguanas in captivity has been limited. Many successful US breedings involve animals that are paired annually or separated only for brief periods. In addition, animals that have been raised together are much easier to pair than animals that are paired for the first time as adults. Captive iguanas vary in disposition and some species (e.g., Grand Cayman Blue iguanas) can be very aggressive toward conspecifics outside the breeding season. Most of the taxa kept in captivity can be raised together and later kept in pairs; however, after females lay, males often become aggressive and pairs may need to be separated. For purposes of captive reproduction, most facilities agree that housing potential mates together for at least most of the year is advisable.

Pairs should be housed in large enclosures that can be divided during periods of aggression. For aggressive species such as Grand Cayman Blue iguanas, or individual animals that are aggressive toward mates during the non-breeding season, chemical and visual contact should be maintained when animals are not housed together if breeding is to be attempted. Plexiglass windows with holes drilled through them, or even small, plastic-coated screens between cage walls, work well in maintaining contact between animals (Figure 4.17). This same type of contact can also be used between cages of adult males to stimulate breeding. When contact between males is desirable to stimulate an interest in breeding behavior but not logistically possible, placement of small mirrors within enclosures can be useful.

In order to reproduce, female iguanas must have sufficient body weight. Weight gain and maintenance is usually not a problem when high basking temperatures and plenty of food are offered. A slight temperature drop combined with a reduction in light cycle for the late winter and spring often help to stimulate breeding when temperature and day length are subsequently increased. In most US facilities, breeding takes place from April to July, depending on species and temperature/light cycles. The majority of copulations occur around June, with oviposition in July.

Copulation may appear to be somewhat rough and females often bear small injuries to the nuchal crest. Over-aggressive males can injure females. If males prevent females from eating,

FIGURE 4.17 A small screen between cages will often help keep separated iguanas in visual and chemical contact to make reintroduction for breeding easier and allow males to detect when females are cycling.

constantly chase them, bite their crest so severely that it bleeds excessively or loses scales, the animals should be separated. For animals that are separated except during breeding, the first introductions can be dangerous for the female, and keepers should monitor the situation closely and be ready to remove over-aggressive males. Males should always be placed into the cages of females for breeding to reduce the risk of a territorial assault on the female. Animals with a strong pair bond usually copulate without excessive roughness, feed and bask together, and males often protect females from keepers. Animals with this type of bond usually do not have to be separated during nesting, although females may become aggressive towards males at this time. In the rare event that females do not seem stressed by the male's presence and continue feeding until just before oviposition, the male should be allowed to stay in the enclosure. In most cases, males should be removed from the female's enclosure for nesting.

With more and more instances of successful copulation in captivity, nesting is emerging as the primary roadblock to successfully breeding rock iguanas. In many facilities, reports of successful copulations and fertile eggs are confounded by low hatching rates. It seems that even with fertile eggs, if a female does not nest properly or holds the eggs too long because she is not comfortable with the nesting situation, the majority of the eggs will not hatch (Figure 4.18). Proper nesting is defined as the female digging a burrow a few days to a few hours before laying, laying the eggs in a timely matter (a few hours, not days), covering the nest, and often defending it (Lemm et al., 2005). Other factors potentially contributing to nesting failures include nutrition and incubation practices.

To achieve nesting success, nesting areas should be deep, spacious, and warm (85–87 °F/29.4–30.6 °C) with optimal soil conditions. Ideally, the substrate throughout the cage should be at least three feet deep so females may choose to nest in a variety of places. At many facilities, especially off-display areas, this depth of substrate is not

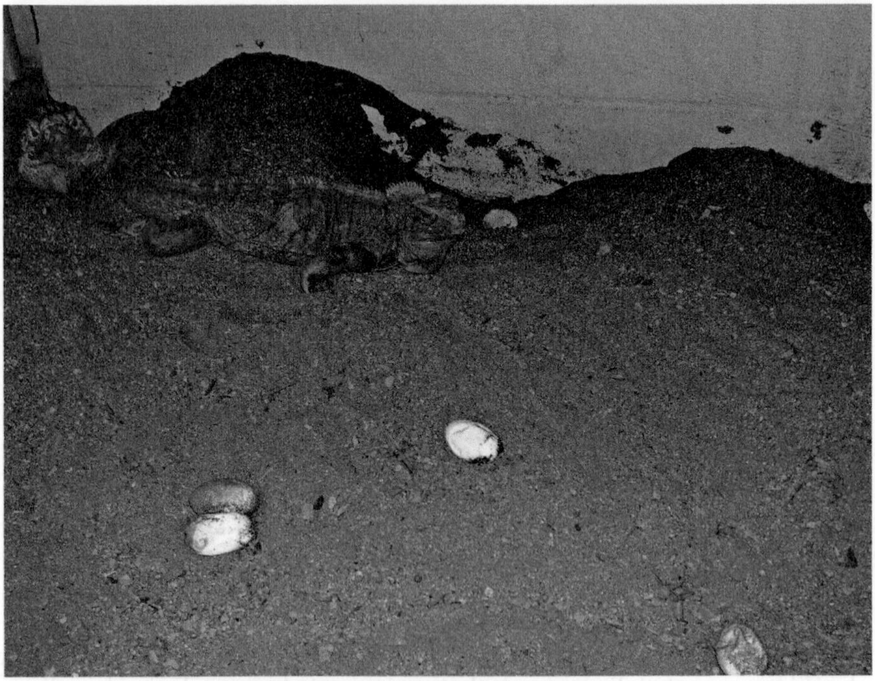

FIGURE 4.18 Unsuccessful Jamaican iguana nest. The female was not comfortable with the nesting situation, held the eggs too long, then discarded them on the soil surface.

possible. In this case, built-in nest boxes, constructed to be as large as possible, should be used (Figure 4.19). Nest boxes measuring 4 ft × 4 ft × 3 ft high (1.2 m × 1.2 m × 0.9 m) have been used with some success. Further, plastic nest boxes, made from large Rubbermaid® tubs or other containers, have worked for some species. However, these nest boxes are usually only successfully used by young animals or first-time breeders for a year or two, followed by below-average nesting success. The Indianapolis Zoo has had nesting success with Rubbermaid® tubs that are completely buried in sand substrate. Animals enter these tubs by digging down and entering a hole cut in the side of the tub (R. Reams, personal communication). Some animals may be quite ingenious in their choice of nesting site (see Figure 4.20).

Some institutions have used sand, rich soil, or a mixture of both with success. Nesting soil should be free of rocks, gravel, and other large debris. Sifted dirt is ideal as it holds moisture, which in turn keeps burrows from collapsing. Nesting areas should be misted whenever they become dry. For more reluctant nesters, items such as rocks or logs may be necessary for animals to burrow underneath. Rock iguanas typically only lay one clutch of eggs per year, and some species may skip a year of reproduction in the wild (Iverson et al., 2004). Multiple-clutching is fairly common in captivity. At the San Diego Zoo, we have seen female Cuban iguanas lay up to three clutches in a single season (April–October), usually spaced 2–2.5 months apart. On average, smaller females at our facility usually only produce one clutch per year.

FIGURE 4.19 Nest boxes can be used successfully with young iguanas, but as they age, iguanas usually do not nest successfully in constrained areas.

After oviposition, females that have laid fertile eggs often guard the nest. Occasionally those that have laid infertile clutches will guard nests as well, but to a lesser degree. In most species it is easy to tell when a female has laid because the burrow has been closed and the female looks thin. However, females of some species, such as the Jamaican iguana, tend to maintain their shape after laying and nests need to be carefully inspected. If the female aggressively defends the nest, she should be captured briefly until the eggs are collected and the nest can be excavated and refilled. Finding eggs can be very difficult, and care must be taken when digging, especially in corners where iguanas often deposit eggs. Eggs should never be rotated from the position in which they were found, and in order to keep track of individual eggs, numbers can be written directly on the eggshell with a pencil. Eggs should be carefully weighed and measured and placed in incubators.

Perlite or vermiculite should be used as incubation medium. The mass of substrate should equal 3–5 times the mass of the total clutch of eggs, and should be placed in a large plastic box with a loose-fitting lid. It is important that ample air space is provided above the eggs within the box. An optimal incubation box should have as much air space as substrate. Substrate is mixed with water at a ratio of 1:1, by weight, and incubated at 84.2–87.8°F (29.0–31°C). Wild nest temperatures for Jamaican and Anegada iguanas fluctuate between

FIGURE 4.20 Successful Grand Cayman Blue iguana nest in a potted *Ficus* plant.

83.3 and 92.3 °F (28.5–33.5 °C) and 83.1 and 91.2 °F (28.4–32.9 °C), respectively. While successful incubation has been recorded at both extremes of these ranges, the protocol recommended here represents the mean at which hatchlings can be expected to emerge with their yolk sacs close to fully absorbed.

A 1 : 1 ratio of substrate to water is used in most facilities and is valuable in situations where eggs are slightly desiccated (usually because eggs were not found in a timely manner or nesting substrate was too dry). Using the 1 : 1 ratio, egg boxes rarely need to have water added to them. Some facilities use a Perlite/vermiculite to water ratio of 2 : 1. This ratio is most often used in facilities that weigh egg boxes regularly and add lost water upon each weighing. The 2 : 1 ratio is also useful in instances when the incubator in use does not have a fan or egg boxes are not well ventilated, because the drier mixture helps keep excess moisture from building up on the inside of egg box lids and dripping on the eggs.

Eggs should be placed on top of the substrate rather than buried (Figure 4.21), and should not be in contact with each other. Some facilities choose to weigh egg boxes weekly

FIGURE 4.21 Incubation setup for rock iguana eggs at the Griffin Reptile Conservation Center. Perlite is used as a substrate.

and add water that is lost over time. This works well up until the last few weeks of incubation, but increasing water at this time may kill the embryos. Generally, water replacement is done during the first two-thirds of the incubation period only. Some institutions add small amounts of water to the substrate if the eggs start to look too dry. When using this method, warm water can be sprayed between eggs (never directly on them) with a spray bottle, or eggs can be removed and the substrate sprayed evenly. An optimal moisture level will result in the substrate staying clumped when squeezed by hand.

Egg boxes should have loose-fitting lids that allow for minimal air exchange and are removed briefly about every 1.5–3 weeks during the first two months of incubation for additional air exchange. Near the end of incubation (70–128 days, depending on species and incubation temperature), lids should be removed daily to allow for the increased respiration rates of developing embryos. Water is never added during the last trimester of incubation, when eggs begin to lose weight and wrinkle prior to hatching. Some eggs may start to develop mold spots even when fertile. Gently wiping spots from egg surfaces with a cotton swab and 1% iodine solution is an effective method for mold removal.

If breeding, nesting, and incubation were successful, hatching success is usually quite high (Figures 4.22 and 4.23). When full-term, apparently healthy young fail to pip or pip and die, a probable cause of death is adding too much water to the egg boxes during the final weeks of incubation. Hatchlings commonly remain in the egg for up to a day prior to full emergence with just their heads visible (Figure 4.24).

HATCHLING CARE

Hatchling iguanas should be removed from incubation boxes immediately and placed in small tubs on clean, moistened paper towels (Figure 4.25). New hatchlings should be housed individually as they can easily injure one another. Boxes can then be placed back in the incubator until the umbilical scars close and heal, and yolk sacs are completely absorbed.

Swollen or bleeding yolk sacs/umbilical regions should receive immediate veterinary care and may need to be removed manually if they do not heal properly. Once umbilici are healed, hatchlings can be housed separately in 10–20 gallon terraria or housed as a group in larger enclosures. When housed separately in terraria, most soil substrates

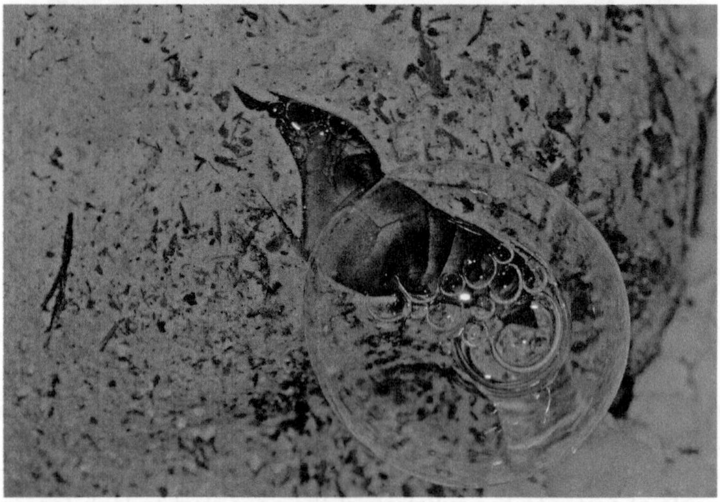

FIGURE 4.22 A Grand Cayman Blue iguana's first breath.

FIGURE 4.23 Hatchlings from the first known captive breeding of the Anegada Island iguana.

work well in conjunction with small plants and a small water dish (Figures 4.26 and 4.27). A temperature gradient of 75– 85 °F (23.9– 29.4 °C) should be maintained with a hide area on both sides of the gradient. Hatchlings should be misted 3–5 times a week, as they may not recognize water dishes at this age. It is also advisable to keep a small portion of the

FIGURE 4.24 Grand Cayman Blue iguana in the "sleeping bag" phase of hatching. Iguanas often stay in this position for up to a day or two before completely hatching out of the egg.

FIGURE 4.25 Freshly hatched iguanas such as this Grand Cayman Blue iguana should be kept in the incubator on a moist paper towel until the umbilical area heals.

FIGURE 4.26 Cages for newly hatched iguanas at the San Diego Zoo.

substrate moist at all times as young iguanas dehydrate quickly when basking at high temperatures, especially if they have not yet started eating large amounts. A suitable basking site up to 120 °F (48.9 °C) surface temperature is sufficient. Access to ultraviolet light is critically important for proper bone mineralization in young iguanas. Hatchlings should be fed every day (Table 4.2).

Within the first month of hatching, the gender of individual animals can be determined with sexing probes. These small, metal probes are lubricated and gently inserted into the cloaca, pointing toward the posterior of the tail. This is a delicate procedure that should be performed by experienced keepers and veterinarians because injury to the male reproductive organs (hemipenes) is possible (Figure 4.28). The probe depth is greater in males at the point where the probe enters the inverted sacs containing the hemipenes. Probe depths vary among species, although species of similar body size usually have comparable probe depths. In the smallest species, the Turks and Caicos iguana, probe depths of hatchling males and females are 10+ mm and 4+ mm, respectively. In adults, probe depths are 25+ mm for males and under 15 mm for females. In larger species such as the Cuban iguana, hatchling male and female probe depths measure 15–20 mm and around 4–8 mm, respectively. Adult probe depths range from 25 to 40 mm for males and 12–15 mm for females. Because sex determination is genetic in rock iguanas, sex ratios within clutches should be near unity.

Passive integrated transponder (PIT) tags for animal identification can be inserted into the left rear thigh of hatchlings of most species as early as one month after hatching (Figure 4.29).

FIGURE 4.27 Hatchling and juvenile enclosures at the Blue Iguana Recovery Program breeding and headstart facility.

In the past, PIT tags were inserted into many locations, including the legs, body wall, and neck; however, the IUCN Iguana Specialist Group recommends following the established convention of using the left rear thigh as an insertion site. Because of the risk of injury to the animal, PIT tag insertion should only be undertaken by trained personnel.

When housed in groups, hatchling iguanas establish dominance hierarchies relatively quickly. Numerous and spatially dispersed food bowls, basking sites, and retreats should be provided as soon as larger animals begin to show evidence of aggressive behavior. Separate enclosures may be needed for smaller hatchlings that appear thin or stressed, or constantly hide. Within the first year of hatching, future mates can be housed together. At the San Diego Zoo, large, plastic vegetable bins (RK2 Systems, Inc.) measuring 4 ft × 4 ft × 3 ft (1.2 m × 1.2 m × 0.9 m) are utilized to house up to three yearlings. For more aggressive species such as the Grand Cayman Blue iguana, hatchlings may need to be housed individually. Bins are covered with a wood frame consisting of 0.5 in × 1.0 in mesh. Hatchlings and juveniles are housed indoors and basking heat and ultraviolet light are provided by 275 Watt Active UV Heat bulbs (T-Rex, San Diego, CA) (Figure 4.30). Natural light is provided through UV-transmitting plastic skylights. Iguanas can also be moved outside to temporary sunning cages, provided they have shade at all times as well as basking areas. Dirt is used as a substrate at a sufficient depth to allow the animals to burrow and PVC tubing is provided as refugia. At 2–3 years of age, juveniles are transferred to adult enclosures.

FIGURE 4.28 Probing a hatchling Turks and Caicos iguana to determine gender. Probing should be carried out by experienced keepers or veterinarians to avoid injury to the animal.

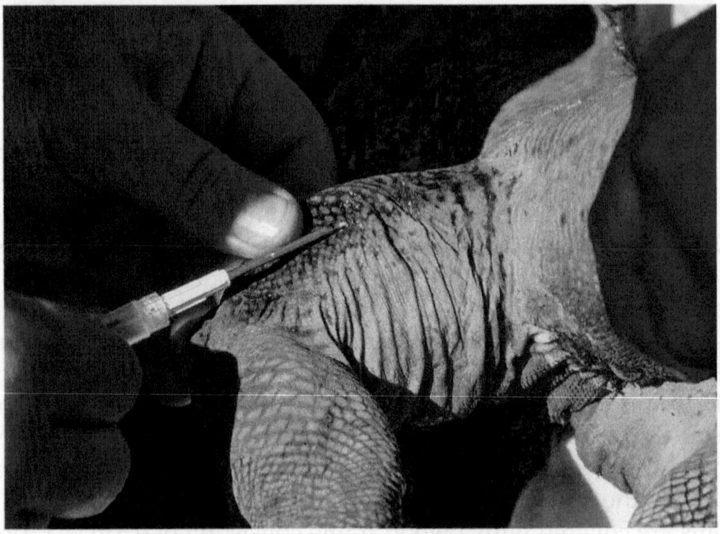

FIGURE 4.29 A PIT tag is being placed into the thigh of this Anegada Island iguana in order to identify it with a distinct alphanumeric code.

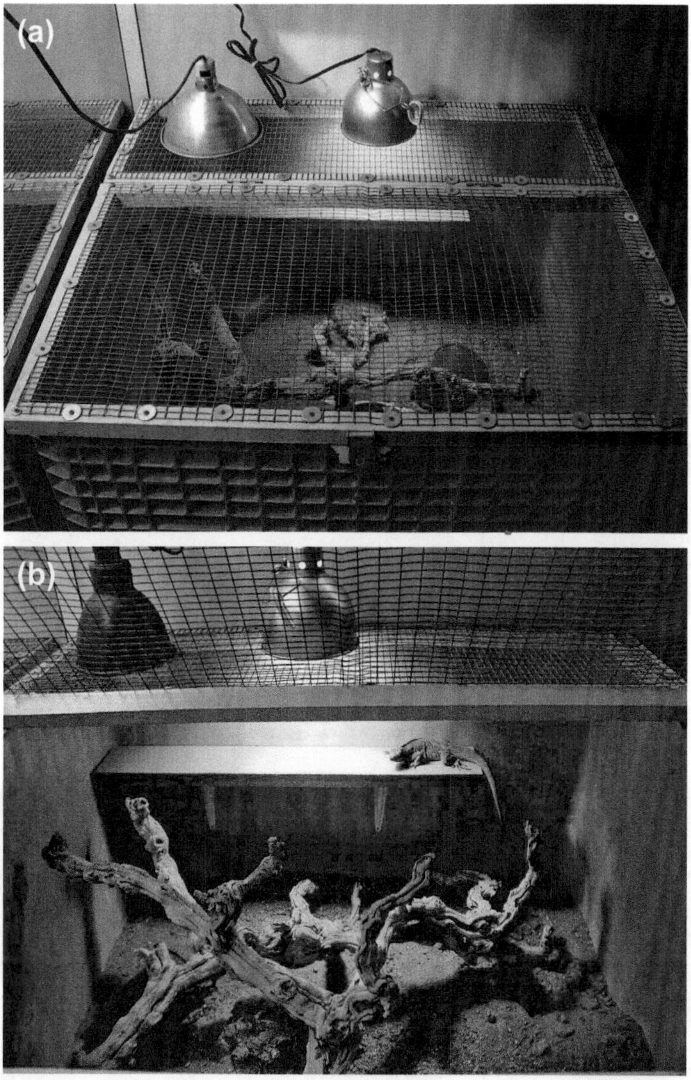

FIGURE 4.30 Hatchling (*a*) and juvenile (*b*) enclosures at the Griffin Reptile Conservation Center.

RECORD KEEPING

A comprehensive database should be kept for all species of West Indian iguanas in captivity. Records of feeding, diet changes, temperature, humidity, captures, animal moves, copulation, egg-laying, veterinary treatments, and any other important notes should be kept on a daily checklist that can be later entered into a computer database.

Because some species have been held in captivity for only a short time, growth data are very important for husbandry and research purposes. Hatchlings, juveniles, and adults

FIGURE 4.31 Measuring the head length of a Turks and Caicos iguana using the lateral measurement. Some researchers use the top of the head measurement or both of these measurements.

FIGURE 4.32 Measuring the snout-to-vent length (SVL) of a juvenile Anegada Island iguana.

FIGURE 4.33 Captive iguanas such as "Gitmo," this large Cuban iguana, should be weighed regularly.

should be weighed and measured at least once a month. Optimal sets of measurements include head length and width (mm), jowl or head width (mm), SVL (mm), tail length (mm), and mass (g or kg) (Figures 4.31–4.33). For older animals that are easily stressed by capture, some facilities choose to forgo measurements during the breeding season. At a minimum, it is important to maintain a record of female body mass before and after egg laying in order to monitor health, as some female iguanas may lose up to a third of their body mass following oviposition. Other notes on captive reproduction are useful as well, including parentage, notes on nesting, egg mass, egg length and width, incubation media, temperature and moisture levels during incubation, length of incubation (oviposition to hatching), and hatchling measurements. Some facilities and researchers also note the probe-depth when sexing hatchling iguanas.

CHAPTER

5

Nutrition

CHAPTER

5

Nutrition

Ann M. Ward, Janet L. Dempsey

OUTLINE

Introduction	129
Feeding Ecology and Digestive Morphology	129
Nutrient Content of the Natural Diet	132
Nutrient Requirements	132
Food Availability and Practical Captive Diets	136
Seasonal Changes	139
Nutrition-Related Health Concerns	140
Vitamin D Needs and Assessing Vitamin D Status	141
Serum Vitamins	142

INTRODUCTION

The West Indian rock iguanas are considered to be the most critically endangered group of lizards in the world (Alberts, 2000). The nutritional needs of these animals must be adequately met to maintain health and reproduction in captivity. Captive diets should be designed to best fulfill nutrient requirements, while also contributing to normal feeding behavior. Key factors to take into account when developing dietary guidelines include information on feeding ecology and digestive morphology, published data on nutrient requirements, and food preferences and availability.

FEEDING ECOLOGY AND DIGESTIVE MORPHOLOGY

Comprehensive nutritional studies describing the foods consumed by wild rock iguanas across seasons, including their contribution to the overall diet and chemical nutrient analysis, have yet to be undertaken. Investigators have described portions of the diets of rock iguanas

Cyclura, First Edition DOI: 10.1016/B978-1-4377-3516-1.10005-6

FIGURE 5.1 Hatchling Turks and Caicos iguana with a mouth stained from eating the fruit of a Sea Grape tree (*Coccoloba* sp.).

by observing animals foraging in the wild or by identifying diet items through examination of stomach contents or fecal remains (Figures 5.1, 5.2). Because of methodological limitations, it is important that these data are interpreted with care (Table 5.1).

Stomach contents do not provide a complete assessment of diet due to differences in the rate of digestion and selective retention of different foods in the gastrointestinal tract (Baer, 2003). Although seeds in fecal remains may reflect fruits consumed, well-digested leaves often are difficult to identify and some vegetation and soft-bodied insects may be completely digested, leaving no remains (Auffenberg, 1982; Jordan, 2005). The most comprehensive work to date continues to be Auffenberg's (1982) review of the Turks and Caicos rock iguana, which describes diet based on stomach contents and fecal analysis, including chemical analysis of the energy content of food items. Taken together, these observations suggest that rock iguanas are predominately herbivorous, with some species consuming animal matter opportunistically. In addition, there are little or no data indicating that the diet of hatchlings and juveniles differs significantly from that of adults. Quantitative data are still needed to accurately define the diets of rock iguanas in the wild.

Examination of gastrointestinal tracts indicates that rock iguanas have a capacious hindgut that is well adapted for their primarily herbivorous diet (Iverson, 1982). The hindgut is distinctive, with several valves in the colon to slow passage rate and increase surface area. Among species studied to date, the Exuma Island and Rhinoceros iguanas have the largest number of valves in their hindgut. These taxa also have few, if any, field reports of

TABLE 5.1 Forage and Food Types Consumed by Free-ranging West Indian Rock Iguanas[1]

Cyclura species	Plant matter				Animal matter
	Leaves	Fruits	Flowers	Other	
Turks & Caicos iguana (*Cyclura carinata carinata*)[a]	X	X	X	X	X
Bartch's iguana (*Cyclura carinata bartschi*)[b]	X	X	X		X
Jamaican iguana (*Cyclura collei*)[c]	X	X	X		X
Rhinoceros iguana (*Cyclura cornuta cornuta*)[d]	X	X	X		
Mona Island iguana (*Cyclura cornuta stejnegeri*)[e]	X	X			X
Andros Island iguana (*Cyclura cychlura cychlura*)[f]	X	X	X		
Exuma Island iguana (*Cyclura cychlura figginsi*)[g]	X	X	X		
Allen Cays iguana (*Cyclura cychlura inornata*)[h]	X	X	X		X
Cuban iguana (*Cyclura nubila nubila*)[i]	X	X	X		X
Sister Isles iguana (*Cyclura nubila caymanensis*)[j]	X	X	X		X
Grand Cayman Blue iguana (*Cyclura lewisi*)[k]	X	X	X	X	X
Anegada Island iguana (*Cyclura pinguis*)[l]	X	X	X		X
Ricord's iguana (*Cyclura ricordii*)[m]	X	X	X		X
San Salvador iguana (*Cyclura rileyi rileyi*)[n]	X	X	X		X
White Cay iguana (*Cyclura rileyi cristata*)[o]	n/a	n/a	n/a	n/a	n/a
Acklin's iguana (*Cyclura rileyi nuchalis*)[o]	n/a	n/a	n/a	n/a	n/a

[1]*These data gathered from published reports using direct observations, analysis of stomach contents, analysis of colon contents, and fecal samples.*
[a]*Auffenberg, 1982. Other plant matter includes buds and mushrooms. Animal matter includes opportunistic ingestion of insects, crustaceans, rodents, fish, and birds.*
[b]*Buckner and Blair, 2000b. Animal matter includes opportunistic ingestion of insects, mollusks, crustaceans, arachnids, lizards and carrion.*
[c]*Vogel, 2000. Animal matter includes opportunistic ingestion of snails.*
[d]*Ottenwalder, 2000a.*
[e]*Wiewandt and Garcia, 2000. Animal matter includes opportunistic ingestion of caterpillars.*
[f]*Buckner and Blair, 2000a.*
[g]*Knapp, 2000a. Reference includes the observation that this species may be coprophagous.*
[h]*Iverson, 2000. Animal matter includes opportunistic ingestion of crabs.*
[i]*Perera, 2000. Animal matter includes opportunistic ingestion of crabs.*
[j]*Gerber, 2000a. Animal matter includes opportunistic ingestion of land crabs and slow moving insect (i.e., Lepidopteran) larvae.*
[k]*Burton, 2000. Other plant matter includes opportunistic ingestion of fungi. Animal matter includes opportunistic ingestion of crabs and cicadas.*
[l]*Mitchell, 2000; Gerber, 2000b. Animal matter includes opportunistic ingestion of some insects at a very low level (<1% of diet as consumed).*
[m]*Ottenwalder, 2000b. Animal matter includes opportunistic ingestion of insects and crustaceans.*
[n]*Cyril et al, 2001; Hayes, 2000b.*
[o]*No published data to date on this species.*

opportunistic ingestion of animal material. Presumably, these gut adaptations facilitate a symbiotic relationship with a microbial population within the gut similar to other herbivorous animals. A significant benefit of this relationship is the production of volatile fatty acids by the microbes, which can be absorbed by the host to provide a significant source of energy. Although few data specific to rock iguanas exist, Baer (2003) provides a comprehensive review of adaptations to herbivory, including important dietary components, digestion, microbial contributions, and fermentation in green iguanas *(Iguana iguana)* that provides a basis for understanding the primarily herbivorous rock iguanas.

NUTRIENT CONTENT OF THE NATURAL DIET

Table 5.2 summarizes available data on the nutrient analysis of native plant parts consumed by free-ranging rock iguanas. Reported dry matter content varies greatly, possibly due to incomplete seed removal for some samples resulting in higher values for many nutrients than are actually utilized by iguanas. These data suggest that the fruits consumed are low to moderate in protein, containing less than 9% protein on a dry matter basis (DM), while leaves and flowers are moderate to good sources of protein at 9–17% DM. Although samples sizes are limited and large variation exists, it appears that flowers contain levels of protein and fiber similar to leaves. Plant parts such as seeds and skins of fruits and non-digestible leaf fractions, while not completely digested by iguanas, may play a role in maintaining integrity and health of the gastrointestinal tract by virtue of their physical form.

Table 5.2 also compares the nutrient content of native plant parts to data for some commercially available produce commonly used in US captive facilities. Overall, commercially available produce tends to be lower in dry matter, fiber (both acid and neutral detergent fiber fractions), and ash (a measurement of total mineral content). Crude protein content for produce tends to be similar or slightly lower compared to native plants. Browse species commonly used in US captive facilities tend to have dry matter in the same range as native plant parts and are similar in fiber when compared to leaves of native plants.

NUTRIENT REQUIREMENTS

There are currently no published data on nutrient requirements for rock iguanas. Therefore, it is reasonable to consider data for species with similar feeding ecology and gastrointestinal tracts, such as the green iguana. In addition, domestic animals have been studied in great detail and published nutrient requirements for some of these species also provide data for comparison.

Data on herbivores including green iguanas (Allen and Oftedal, 2003), horses (National Research Council, 2007), and rabbits (National Research Council, 1977) have been used to establish recommended dietary nutrient ranges for rock iguanas (Table 5.3). Nutrient requirement information for the omnivorous dog (National Research Council, 2006) was also used, as there are some reports of rock iguanas opportunistically consuming animal material. In general, these levels represent minima for most nutrients. Based on the products available and sample diets

TABLE 5.2 Nutrient Analysis on a Dry Matter (DM) Basis of Energy, Protein, Acid Detergent Fiber (ADF), Neutral Detergent Fiber (NDF), Fat and Ash of Plants Consumed by *Cyclura* spp. In Situ Compared to Foods Available in US Captive Facilities

Species	Part[1]	N	DM (%)	Energy (kcal/g)	Protein (%)	ADF (%)	NDF (%)	Fat (%)	Ash (%)
C. c. carinata[2]	Fr	7	n/a	4.6 ± 0.5	n/a	n/a	n/a	n/a	n/a
	L	10	n/a	4.4 ± 0.2	n/a	n/a	n/a	n/a	n/a
C. collei[3]	Fr	5	34.9 ± 18.0	n/a	5.8 ± 2.7	40.5 ± 20.8	40.3 ± 16.4	n/a	n/a
	Fl	2	23.4 ± 6.8	n/a	12.3 ± 5.4	40.5 ± 27.6	25.6 ± 8.0	n/a	n/a
	L	4	58.5 ± 13.8	n/a	16.2 ± 3.1	20.3 ± 8.8	33.3 ± 14.2	n/a	n/a
	S	1	82.1	n/a	9.6	46.8	69.4	n/a	n/a
C. lewisi[4]	Fr	3	18.5 ± 7.3	n/a	4.5 ± 2.7	19.8 ± 6.3	23.6 ± 8.8	2.5 ± 1.9	10.4 ± 3.1
	Fl	5	15.7 ± 4.5	n/a	16.8 ± 6.5	21.3 ± 5.6	31.2 ± 9.4	3.5 ± 1.4	7.3 ± 2.7
	L	8	17.1 ± 5.1	n/a	16.5 ± 4.4	27.0 ± 7.8	36.8 ± 6.0	3.6 ± 1.5	16.6 ± 12.4
	W	11	23.6 ± 12.5	n/a	15.5 ± 6.1	24.0 ± 9.2	37.6 ± 10.9	3.5 ± 2.0	16.4 ± 10.4
C. pinguis[5]	Fr	9	37.5 ± 11.7	n/a	8.8 ± 5.7	35.1 ± 11.3	40.7 ± 15.0	5.0 ± 5.1	n/a
	Fl	6	50.9 ± 26.8	n/a	9.0 ± 3.4	38.0 ± 13.8	51.5 ± 22.3	5.5 ± 6.9	n/a
	L	16	39.0 ± 12.9	n/a	10.3 ± 4.4	22.5 ± 8.6	30.4 ± 10.5	4.6 ± 2.6	n/a
Commercially available fruit[6]		7	14.3 ± 3.8	n/a	7.1 ± 3.7	8.7 ± 5.7	13.1 ± 6.7	n/a	3.9 ± 2.5
Commercially available vegetables[6]		7	10.3 ± 6.7	n/a	14.5 ± 9.0	12.1 ± 5.0	16.9 ± 3.9	n/a	6.9 ± 2.6
Commercially available leafy green vegetables[6]		6	8.8 ± 3.0	n/a	26.0 ± 8.3	14.5 ± 2.2	18.2 ± 1.8	n/a	11.3 ± 2.3
Readily available browse[7]		3	37.9 ± 16.9	n/a	n/a	23.6 ± 13.6	44.1 ± 24.6	n/a	n/a

[1] *Plant parts: flowers (Fl), fruits (Fr), leaves (L), whole plant (W), seeds (S).*
[2] *Auffenberg, 1982.*
[3] *A. Ward and J. Dempsey, unpublished data. Note seeds were not removed from the fruits prior to analysis.*
[4] *A. Ward and J. Dempsey, unpublished data. Note all seeds were removed from the fruits prior to analysis.*
[5] *A. Ward and J. Dempsey, unpublished data. Note most seeds were removed from the fruits prior to analysis.*
[6] *Schmidt et al. (1999). Note values for fruits and vegetables include seeds and skin; values for leafy green vegetables include stem and stalk. Fruits include apple, banana, blackberries, green grapes, kiwi, orange, and papaya. Vegetables include broccoli, carrot, cucumber, green pepper, acorn squash, yellow squash, and sweet potato. Leafy greens include collard greens, kale, romaine, mustard greens, turnip greens, and celery.*
[7] *A. Ward, unpublished data.*

FIGURE 5.2 Mona Island iguana feeding on fallen fruit.

presented in Table 5.3, most captive diets provide levels exceeding these targets. It is not unusual for many diets currently offered to adults to already meet or exceed the minima for growth.

Recommendations for egg laying were not made due to a lack of species-specific data. For poultry in maximum egg production, calcium levels fed may reach 3% of the diet (National Research Council, 1994). Levels up to 2% should be safe for iguanas. Levels above 2.5% are not recommended because of potential interference with the absorption of other minerals (Klasing, 1998). A study evaluating dietary calcium levels and calcium balance in the insectivorous Drakensberg crag lizard (*Pseudocordylus melanotus melanotus*) demonstrated that calcium balance was maintained at levels ranging from 1.4 to 5.6% dietary calcium on a dry matter basis (van der Wardt, 1999). However, at 29 days the duration of feeding was relatively short, and this species is not necessarily representative of calcium metabolism in predominantly herbivorous lizard species. Long-term feeding and assessment in rock iguanas is necessary to determine if similar dietary calcium levels are safe and/or beneficial. During all physiological states, adequate vitamin D is required for normal calcium metabolism.

A comprehensive review of the role and importance of macro- and micro-nutrients is provided by Allen and Oftedal (2003). In general, quantifiable differences in nutrient requirements for juvenile iguana growth and adult maintenance have not been determined. Investigations of hatchlings and juveniles have identified differences in growth rates based on protein and fiber levels fed (Allen et al., 1989; Donoghue, 1995; Baer et al., 1997). Calcium and phosphorous levels needed to maintain adequate growth in herbivorous and omnivorous mammals are shown in Table 5.3. Currently, no dietary level of vitamin D3 has been shown to avoid vitamin D deficiency. Consequently, all iguanas should be provided with an adequate source of UV light (in the range 295–300 nm), known to support the most effective biogenesis of vitamin D (Holick, 1995).

TABLE 5.3 Composition and Nutrient Content of Example Diets Fed as an Average Over the Year (Diet 1 and 2) or Seasonally (Diet 3 and 4) to Captive Rock Iguanas Compared to Minimum Recommended Nutrient Ranges

	Diets – Percent of the diet as fed			
Food category	*Diet 1*	*Diet 2*	*Diet 3*	*Diet 4*
Nutritionally complete dry herbivore diet	2.8	—	—	—
Nutritionally complete dry primate diet	2.8	19.7	—	—
Nutritionally complete omnivore gel diet	7.2	—	—	—
Leafy greens	80.0	16.9	53.3	76.1
Vegetables	4.8	24.1	22.1	13.9
Fruit	2.4	14.0	22.1	7.5
Water	—	25.3	—	—
Nutritional supplements	—	—	2.5	2.5
Total	100.0	100.0	100.0	100.0

		Diets – Calculated nutrient levels on a dry matter basis				
Nutrients	***Units***	***Minimum recommended dietary nutrient range for* Cyclura *spp.*[a]**	*Diet 1*	*Diet 2*	*Diet 3*	*Diet 4*
Crude protein	%	17.0–26.0 growing 12.0–17.0 maintenance	25.6	15.7	11.4	14.1
Crude fat	%	3.0	4.5	6.6	2.0	2.3
Linoleic acid	%	1.0		n/a	n/a	n/a
Crude fiber	%	—	9.1	104	n/a	n/a
Acid detergent fiber	%	13.0–18.0	n/a	11.1	9.3	10.9
Neutral detergent fiber	%	—	n/a	20.4	14.4	13.9
Vitamin A[c]	IU/kg	5000	4952	16377	—[c]	—
Vitamin D	IU/kg	—	1379	2870	—	—
Vitamin E	IU/kg	150.0	180.6	243.1	67.2	83.1
Vitamin C	mg/kg	—	n/a	789.2	2582	3375
Vitamin K	mg/kg	1.0	n/a	n/a	n/a	n/a
Thiamine	mg/kg	8.0	8.7	12.5	4.7	4.6
Riboflavin	mg/kg	5.0	11.7	12.7	5.6	6.9
Pantothenic acid	mg/kg	15.0	16.9	62.2	19.5	21.0
Niacin	mg/kg	90.0	69.4	102.5	43.8	48.7
Pyridoxine	mg/kg	6.0	n/a	15.9	11.5	11.7

(*Continued*)

TABLE 5.3 Composition and Nutrient Content of Example Diets Fed as an Average Over the Year (Diet 1 and 2) or Seasonally (Diet 3 and 4) to Captive Rock Iguanas Compared to Minimum Recommended Nutrient Ranges—cont'd

		Diets – Calculated nutrient levels on a dry matter basis				
Nutrients	***Units***	***Minimum recommended dietary nutrient range for* Cyclura *spp.*[a]**	***Diet 1***	***Diet 2***	***Diet 3***	***Diet 4***
Folic acid	mg/kg	0.8	n/a	9.6	6.0	7.3
Biotin	mg/kg	0.25	n/a	n/a	n/a	n/a
Vitamin B_{12}	mg/kg	0.03	n/a	n/a	n/a	n/a
Choline	mg/kg	1200	n/a	1530	400.1	591.0
Calcium	%	1.0 growing 0.6 maintenance	1.2	1.1	3.7	4.8
Phosphorus	%	0.8 growing 0.5 maintenance	0.7	0.6	0.2	0.3
Potassium	%	0.5	n/a	1.3	2.0	2.4
Sodium	%	0.2	0.6	0.3	0.2	0.4
Magnesium	%	0.15	0.3	0.17	0.17	0.19
Iron	mg/kg	80.0	217.7	236.9	58.4	72.3
Copper	mg/kg	10.0	n/a	26.5	12.6	16.0
Manganese	mg/kg	50.0	n/a	84.6	23.5	31.1
Zinc	mg/kg	82.0	75.6	122.7	25.5	36.9
Iodine	mg/kg	0.6	n/a	n/a	n/a	n/a
Selenium	mg/kg	0.3	n/a	n/a	n/a	n/a

[a] *Recommended values based on Allen and Oftedal, 2003; National Research Council, 1977 (rabbit), 2006 (dog), 2007 (horse).*
[b] *n/a = values for this nutrient were not available in the database for one or more of the diet ingredients contributing significantly to the diet and therefore were not calculated.*
[c] *Diets 3 and 4 do not contain a source of preformed vitamin A (retinol). It is possible, similar to herbivorous mammals, that rock iguana species can utilize carotenoids for vitamin A synthesis (National Research Council, 2007). However, differences in bioavailability of carotenoids for conversion to retinol between fruits, root vegetables, and leafy vegetables precludes the use of a general correction factor to calculate possible retinol contribution (Carrillo-Lopez et al., 2010).*

FOOD AVAILABILITY AND PRACTICAL CAPTIVE DIETS

Table 5.4 outlines composition and nutrient content of diets fed to rock iguanas with repeated reproductive success, and compares them with recommended minimum dietary nutrient ranges. Diets based on nutritionally complete feeds require little or no additional vitamin and mineral supplementation. Table 5.4 lists the nutrient content of some nutritionally complete feeds that are commercially available to US captive facilities.

TABLE 5.4 The Nutrient Content of Some Commercially Available, Nutritionally Complete Feeds on a Dry Matter Basis

		Level in the diet					
Nutrient	**Units**	**A[1]**	**B[2]**	**C[3]**	**D[4]**	**E[5]**	**F[6]**
Metabolizable energy	Kcal/g	1.6	1.7	2.1	1.8	1.6	1.8
Protein	%	26.8	21.4	28.3	25.5	18.3	25.6
Fat	%	3.4	5.1	7.3	4.0	2.7	5.1
Linoleic acid	%	1.2	1.9		2.1	1.6	3.1
Crude fiber	%	17.8	9.9	10.5	9.2	9.1	12.2
Acid detergent fiber (ADF)	%	23.9	13.5	15.8	12.4	13.1	18.9
Neutral detergent fiber (NDF)	%	32.1	26.7	25.1	20.8	23.1	25.7
Ash	%	10.6	9.3	8.6	6.9	7.9	7.7
Calcium	%	2.8	1.3	1.1	1.2	1.5	2.1
Phosphorus	%	1.3	0.8	0.7	0.7	0.6	1.0
Potassium	%	1.9	0.6	0.7	0.4	0.4	1.3
Sodium	%	0.2	0.3	0.1	0.1	0.3	0.5
Magnesium	%	0.3	0.2	0.3	0.2	0.2	0.3
Iron	ppm	614.6	232.4	187.8	277.2	280.5	725.0
Copper	ppm	16.0	28.1	14.0	21.4	29.3	27.6
Manganese	ppm	131.5	233.6	83.1	56.5	84.3	166.7
Zinc	ppm	138.7	173.5	76.9	104.8	172.0	169.7
Iodine	ppm	1.7	1.1	—	1.1	1.4	1.9
Selenium	ppm	0.4	1.5	0.4	0.5	0.3	0.3
Vitamin A	IU/kg	22000	32900	—	8421	44000	7150
Vitamin D3	IU/kg	9130	2200	—	1684	7260	1320
Vitamin E	IU/kg	220.0	220.0	—	210.5	72.6	363
Vitamin K	ppm	4.5	4.4	—	—	3.5	4.0
Vitamin C	ppm	308.0	242.0	—	210.5	550.0	—
Thiamin	ppm	15.4	8.3	—	5.3	9.8	12.1
Riboflavin	ppm	16.5	13.2	—	6.3	9.2	11.0
Pantothenic acid	ppm	56.1	53.9	—	21.1	66.0	35.2
Niacin	ppm	106.7	194.7	—	42.1	132.0	99.0
B6	ppm	15.4	15.4	—	4.2	15.4	5.1

(Continued)

TABLE 5.4 The Nutrient Content of Some Commercially Available, Nutritionally Complete Feeds on a Dry Matter Basis—cont'd

		Level in the diet					
Nutrient	Units	A[1]	B[2]	C[3]	D[4]	E[5]	F[6]
Biotin	ppm	0.5	0.6	—	0.2	0.11	0.6
Folic Acid	ppm	10.2	7.7	—	0.5	8.7	1.9
B12	μg/kg	123.2	31.9	—	21.0	24.2	25.3
Choline	ppm	1837	3410	—	—	1320	1782

[1]*Mazuri Iguana (linoleic acid, iodine, and vitamins provided by the manufacturer, ME, calculated, all other values chemically analyzed). Mazuri/PMI Nutrition International, St Louis, MO.*
[2]*Zeigler Iguana (linoleic acid, iodine, and vitamins provided by the manufacturer, ME, calculated, all other values chemically analyzed). Zeigler Bros. Inc., PO Box 95, Gardners, PA.*
[3]*ZooMed Juvenile Iguana (linoleic acid, iodine, and vitamins provided by the manufacturer, ME, calculated, all other values chemically analyzed). Zoo Med Laboratories, Inc., 3650 Sacramento Dr., San Luis Obispo, CA.*
[4]*Marion Leafeater (linoleic acid, iodine, and vitamins provided by the manufacturer, ME, calculated, all other values chemically analyzed). Marion Zoological, 2003 E. Center Circle, Plymouth, MN.*
[5]*Purina Monkey Diet Jumbo 5037 (linoleic acid, iodine, and vitamins provided by the manufacturer, ME, calculated, all other values chemically analyzed). Mazuri/PMI Nutrition International, St Louis, MO.*
[6]*Mazuri Herbivore ADF16 (all values are approximate provided by the manufacturer). Mazuri/PMI Nutrition International, St Louis, MO.*

Supplementing one or more nutrients can result in toxic levels and/or imbalances if not calculated for in the development of the diet. See Allen and Oftedal (2003) for an in-depth discussion on use of supplements.

An important consideration when including fresh produce in captive diets is the difference in nutrient content when compared to native plant parts (see Table 5.2). In general, cultivated produce is lower in most nutrients than native plants, which supports the inclusion of a nutritionally complete feed in order to meet recommended dietary nutrient levels. Recognizing that some species occasionally consume animal matter, animal products such as gel diets or insects may be incorporated at 5% (no more than 10%) of the diets of these predominately herbivorous animals. In addition, fresh plants in the form of flowers or browse in exhibits can be included, but at no more than 5% by weight as fed. Diets can be altered seasonally, provided recommended nutrient levels are met.

Several publications have addressed concerns regarding oxalates, phytates, and glucosinilates in produce items (Allen and Oftedal, 2003; Donoghue, 2006). In general, problems can be avoided if a varied diet is offered, avoiding one or a few food items contributing significantly. Inclusion of produce often encourages animals to consume nutritionally complete products. Grinding, grating, and/or softening the complete feed and coating the produce can facilitate consumption. Thorough mixing of diet ingredients may facilitate consumption of the total diet by making it difficult for iguanas to select only favored items.

Food items for adults, juveniles, and hatchlings should be chopped to an appropriate size for ease of eating. Multiple food pans may be required to minimize competition among animals fed in pairs or groups. Additionally, visual barriers can be utilized. Consumption of the diet should be recorded and assessed based on the animal's body condition, physiological state, and food items or total food remaining. Rock iguanas should be offered food daily.

SEASONAL CHANGES

Few data exist quantifying changes in nutrient intake with season (wet/dry; non-breeding/breeding) in free-ranging iguanas. Investigations to date are limited to green iguanas. These studies indicate significant variation in metabolizable energy intakes of free-ranging animals within seasons, making between season comparisons difficult (van Marken Lichtenbelt et al., 1997). Different types of foods (fruits, flowers, young and old leaves) are consumed by iguanas on Jamaica and Grand Cayman depending on the season (Burton, 2000; Vogel, 2000) (Figure 5.3). Due to the seasonal climate, it is suspected that seasonal variation in food choices occurs among other rock iguanas as well.

Preliminary data from a limited number of captive Grand Cayman Blue iguanas indicate little, if any, difference between breeding and non-breeding seasons in crude protein and acid detergent fiber intakes on a gram per day per kilogram body weight basis, although diets offered varied in nutrient content (A. Ward, unpublished data). On a dry matter basis, the protein content of the diet consumed at the headstart facility during the breeding and

FIGURE 5.3 A keeper prepares native plants for Grand Cayman iguanas at the Blue Iguana Recovery Program facility.

non-breeding seasons was 9% and 18%, respectively. More research is needed in this area, and should focus on differences in seasonal intake of other nutrients.

NUTRITION-RELATED HEALTH CONCERNS

Metabolic Bone Disease

Several calcium metabolism disorders are grouped under metabolic bone disease. These are well reviewed by Allen and Oftedal (2003), and include nutritional secondary hyperparathyroidism, fibrous osteodystrophy, rickets, osteomalacia, and metastatic mineralization. Metabolic bone disease is a complex group of conditions involving diverse nutrients, metabolites, hormones, and organs with both nutritional and non-nutritional causes. Nutritional causes include insufficient levels of calcium and/or vitamin D (lack of exposure to UV light in the appropriate wavelength range), and imbalances in the calcium to phosphorus ratio in the diet. It is interesting to note that metastatic mineralization in humans occurs as a result of hypervitaminosis D, while some cases of metastatic mineralization in green iguanas appear to be the result of hypovitaminosis D (Richman et al., 1995). Because these problems can be

FIGURE 5.4 A juvenile Grand Cayman Blue iguana with gout. Note the swollen forelimb.

multifactorial, it is important to review the calcium, phosphorus, calcium to phosphorus ratio, and vitamin D levels in the diet, as well as serum parameters, for an accurate diagnosis. Additionally, non-nutritional factors such as renal disease can affect serum parameters. Thus, it is important to review the overall health of the animal.

Gout

High protein diets have often been suggested as a cause of gout (Figure 5.4). However, controlled studies to date in iguanas are few and do not support this theory. As reviewed by Allen and Oftedal (2003), iguana diets containing protein levels as high as 35% did illicit a rise in serum uric acid levels, but these were within the reported normal range for hydrated reptiles (1–8 mg/100 ml). A combination of ingredients fed within the suggested ranges would not exceed an overall protein level of 35%. Considering that elevated serum uric acid levels can reflect a natural post-prandial rise or dehydration or renal insufficiency, values should be interpreted carefully.

VITAMIN D NEEDS AND ASSESSING VITAMIN D STATUS

Meeting Vitamin D Needs

Considering the still common occurrence of metabolic bone disease and studies indicating that normal dietary levels of vitamin D3 do not prevent rickets in captive green iguanas (Bernard et al., 1991), exposure to ultraviolet radiation within the UVB range of 280–315 nm appears to be the most effective method to meet vitamin D requirements. Peak conversion of provitamin D3 to previtamin D3 in the skin of humans occurs in a narrow band between 295 and 300 nm (Holick, 1995). The final conversion of previtamin D3 to vitamin D3 in the skin is temperature-dependent.

UVB Radiation

Outdoor exhibits, exposure to natural sunlight, UVB emitting bulbs, and UVB penetrating skylights are appropriate methods of supplying UVB radiation (Figure 5.5). Methods that supply light in the 295 to 300 nm range are the most effective. Distance of the lamp from the animal, as well as the decrease in UV output over time, must be considered in the placement/replacement of UV bulbs (Bernard, 1995). Most bulbs provide a combination of light, including visible light, UVA radiation, UVB radiation, UVC radiation, infrared radiation, or some combination thereof. Too much, or inappropriate, radiation can result in tissue damage (Gallagher and Lee, 2006). Consequently, all bulbs should be used with caution.

Ultraviolet meters, radiometers, and measuring conversion of previtamin D to vitamin D3 in test ampoules are methods that have been used to determine the effectiveness of bulbs and skylights. Instruments that read within the peak conversion range will provide the most valuable assessment of vitamin D synthesizing capacity. Ampoules containing a solution of 7-dehydrocholesterol in n-hexane when exposed to light can be used to measure vitamin D conversion (Holick et al., 1995). A comprehensive review of this topic and the relationship to

FIGURE 5.5 UVB radiation is necessary for the health and well-being of rock iguanas. Enclosures at the San Diego Zoo's Griffin Reptile Conservation Center allow animals access to outdoor yards to bask in natural sunlight.

reptiles is available in Bernard (1995). A review of broadband radiometers and light bulbs can be found in Gehrmann et al. (2004), with additional assessment of bulbs in Schmidt et al. (2006).

Oral Vitamin D Supplementation

Supplements used for a variety of animals are discussed in Ullrey and Bernard (1999). Levels known to be toxic to other species do not appear to have the same affect on iguanas. A large single oral dose (8.5 IU vitamin D3/g of body weight) increased serum 25 hydroxy D3 levels, but could not sustain normal serum levels up to five weeks (Bernard, 1995). The long-term effect of repeated large oral doses has not been studied.

Assessing Vitamin D Status

Regardless of assessment of light, regular serum analysis for 25-hydroxy vitamin D3 should be conducted to ensure that lights/skylights allow sufficient UVB for animals to maintain levels within a normal range year-round. Because it reflects diet and biogenesis over several weeks to months, serum 25-hydroxy D3 is the most valuable metabolite to assess vitamin D status (Holick, 1990).

SERUM VITAMINS

Although serum vitamin levels have been used to assess nutrient status, many are not ideal and have limitations. In humans, serum levels of retinol have been shown to reflect vitamin A status only if very low or very high (Crissey et al., 1999). Serum retinol has been used to note dietary levels consumed, and serum alpha tocopherol can be correlated with liver stores (Crissey et al., 2003). However, different animals of the same species tend to exhibit individually characteristic alpha tocopherol levels (Shrestha et al., 1998).

Published values do not exist for these vitamins or the carotenoids in rock iguanas. Because low levels of carotenoids may be a reflection of poor health status or illness in some zoo animals (Slifka, 1999), establishing normal values on healthy animals should be pursued. Green iguanas appear to absorb and/or accumulate the carotenoids lutein, zeaxanthin, and to some extent, canthaxanthin (Raila et al., 2002). Beta-carotene does not appear in the serum, suggesting that iguanas do not absorb intact beta-carotene. However, similar to birds, they may be able to convert it to vitamin A at the brush border of the intestine. Rock iguanas may absorb and accumulate carotenoids in a similar manner.

In summary, serum vitamin data should be critically reviewed based on the limitations discussed. These data should be one component of a larger assessment that includes diet analysis, diet consumption, and measurement of other health parameters to determine overall nutritional status.

CHAPTER

6

Health and Medical Management

CHAPTER

6

Health and Medical Management

Nancy P. Lung

OUTLINE

Introduction	147	Clinical Techniques	161
Health and Medical Management of Captive Iguanas	147	Health of Free-Ranging and Headstart Rock Iguanas	166
Reproductive Problems	153	Pathology	171

INTRODUCTION

For the past 15 years there has been intensive focus on the conservation and biology of the West Indian rock iguanas. Prior to that, it was necessary to extrapolate from what was known from other iguanid lizards (primarily the green iguana) and reptiles in general when faced with medical and husbandry issues. What has been learned about rock iguanas more recently greatly increases our knowledge of, and resources for, the health management of this group (Alberts et al., 1998; Lung et al., 2002; Fisse et al., 2004; Gerber et al., 2004, 2006; Raphael, 2004, 2006; Wilson et al., 2004a; Ramer et al., 2005, 2009; James et al., 2006; Maria et al., 2007; Zachariah et al., 2009). This chapter summarizes the medical issues facing rock iguanas in captivity, reviews the medical data that have been collected from free-ranging animals as well as those housed in headstart facilities in range countries, and provides a resource that can be utilized by personnel responsible for health management.

HEALTH AND MEDICAL MANAGEMENT OF CAPTIVE IGUANAS

Captive rock iguanas are maintained under a variety of conditions in zoos, breeding centers, and private collections. In order to maintain good health and achieve reproductive success, there are environmental, nutritional, and social needs that must be met as part of

Cyclura, *First Edition* DOI: 10.1016/B978-1-4377-3516-1.10006-8

TABLE 6.1 Results of a Medical Survey of 28 Zoological Institutions Holding Captive Rock Iguanas

Number of animal records reviewed	380
C. cornuta cornuta	188
C. cornuta stejnegeri	3
C. nubila nubila	91
C. lewisi	66
C. pinguis	8
C. cychlura figginsi	7
C. collei	12
C. ricordii	5
Number of medical or necropsy entries	978
Trauma	118
Parasites	114
Infections	89
Reproductive events	50
Renal failure	34
Calcium/Vitamin D related problems	32
Anorexia/lethargy	22
Intestinal impaction	10
Hypothermia	9
Burns	7
Cystic calculi	1

any effective preventive health program. Knowledge of the common medical conditions seen in rock iguanas is essential to designing preventive health programs specific to each captive collection. A survey of medical records of captive rock iguanas reveals the most common reasons for medical intervention with these animals in captivity (Table 6.1).

Trauma

The most common medical problem seen in captive rock iguanas is trauma (Figure 6.1). Trauma is most commonly induced by conspecifics during aggressive social encounters, but can also occur due to injuries in exhibits and holding areas, or during capture and handling. Overcrowding increases the incidence of conspecific aggression. A caretaker with keen observation skills can identify groupings at risk for aggression and re-adjust the group composition to include more compatible individuals.

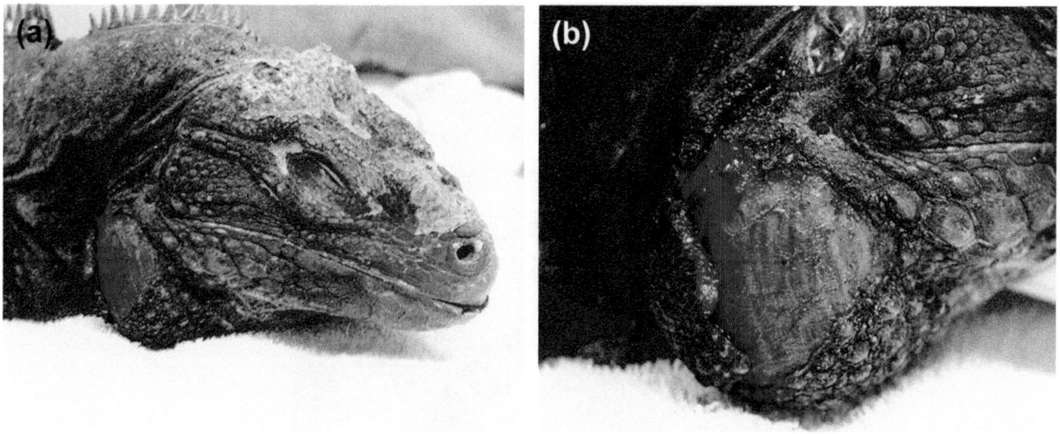

FIGURE 6.1 (*a*) This male Jamaican iguana suffered a degloving injury on both sides of his face during a conflict with another adult male. The close-up photo of the injury (*b*) shows the underlying facial muscles exposed. *Photos by Nancy Lung.*

Lacerations from bite wounds are common. If they are attended to the same day, lacerations can be cleaned and sutured, giving the best outcome. Wounds that are identified more than several hours after occurrence need to be managed as open wounds. This involves thorough cleaning and debridement, as it is common for such wounds to be packed with dirt and other debris. Systemic antibiotics should be used to control infection at the site of the injury. Wounds should be monitored closely, as many develop complications such as abscessation or necrosis of digits or tails. Subsequent amputation of digits, limbs, and tails may be necessary due to infection or loss of blood supply (Figure 6.2).

Long-bone, vertebral, and jaw fractures can also occur during trauma (Figure 6.3). The risk is increased in animals with suboptimal vitamin D and calcium status. Long-bone fractures can be repaired with external splints, external pins, or internal fixation devices (Jacobson, 2003; Mader, 2006).

Parasites

Parasites account for the second most frequently encountered medical event in captive iguanas, and are also seen in free-ranging animals. However, most of these parasites are benign and have not caused clinical problems in otherwise healthy animals. Any parasite can contribute to health problems if animals are in suboptimal weight, physiologically challenged (i.e., gravid or ill), or under environmental stress. In animals maintained on natural substrates, parasite levels can build up in the environment, leading to superinfections and rapid reinfection after treatment. Good hygiene practices and daily removal of fecal material will help with this problem and reduce the need for antihelminthic treatment.

Ticks and mites are the most common ectoparasites. Ticks will attach to the host in a variety of locations, but prefer the more hidden axillary and inguinal regions around the

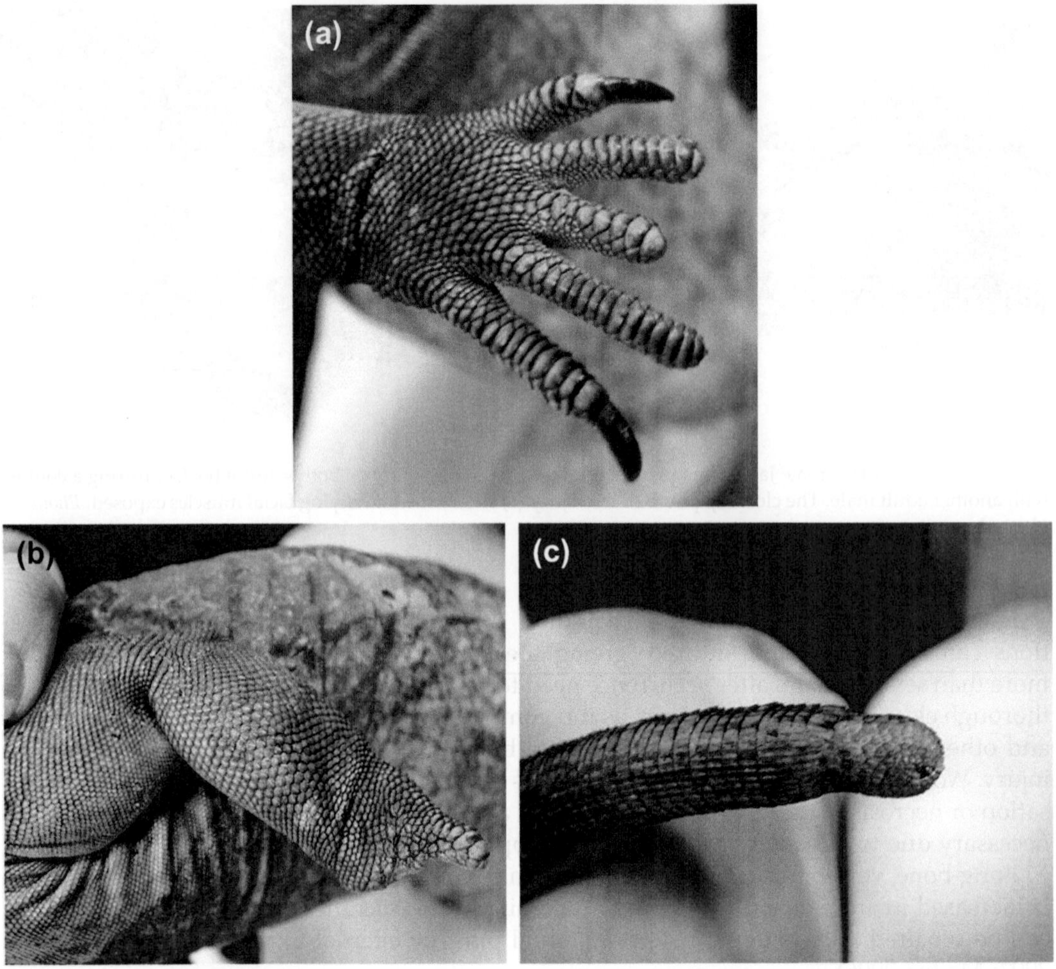

FIGURE 6.2 Loss of nails (*a*), digits (*b*), and tail tips (*c*) is a common consequence of traumatic injuries in rock iguanas. *Photos by Nancy Lung.*

legs, the folds of the cloaca, and occasionally the neck and face. Ticks can be easily missed if the physical exam is not thorough. Ticks rarely cause medical problems for iguanas, although ticks are capable of transmitting hemoparasites and viruses (Telford, 1984; Campbell, 2006). When found they should be removed. Physical removal using forceps is appropriate for low-level infestation. For higher levels of infestation, ticks can be treated topically with a permethrin-based acaracide such as Provent-a-Mite™ (Pro Products, Mahopac, NY) or systemically with ivermectin at a dose of 200 μg/kg body weight (Carpenter, 2005; Plumb, 2008).

Mites are a common ectoparasite of captive iguanas (Figure 6.4). Like ticks, mites prefer folds of skin and can often be seen as red, gray, or black dots around the skin folds of the eyes, neck, cloaca, and limbs. Treatment for mites is the same as for ticks – use of

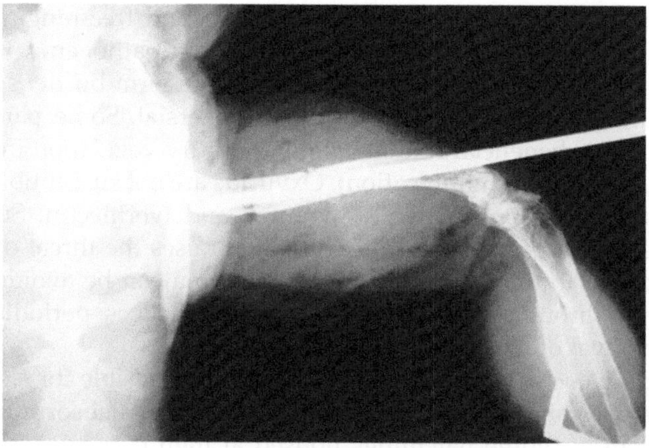

FIGURE 6.3 Repair of a femoral fracture using an intramedullary pin in a Grand Cayman Blue iguana. *Photo by Jan Ramer.*

a permethrin product topically or ivermectin systemically. Environmental control is as important as treatment of the animal and can be done with a cyfluthrin-based spray such as Tempo® (Bayer Corporation, Kansas City, MO).

Endoparasites are diagnosed by fecal exam. Fecal flotation will identify a variety of nematodes (such as pinworms), some cestodes (tapeworms), and some protozoa (including coccidia). A fecal sedimentation is less commonly performed, but is necessary for identifying trematodes. A direct microscopic exam from a fresh sample is used to identify motile protozoa such as ciliates and flagellates. Motile protozoa are generally harmless and are expected to be present in the hindgut of most lizards. A fecal cytology can be performed to identify potentially harmful amoeba.

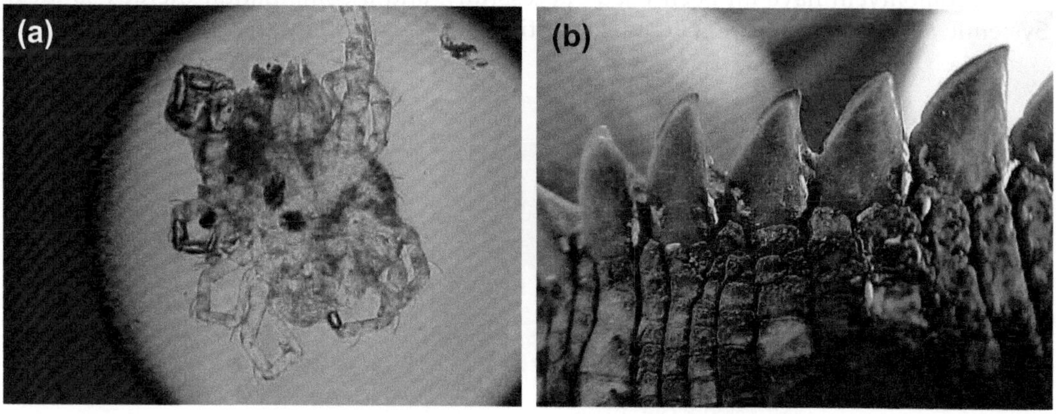

FIGURE 6.4 External mites identified on a Jamaican iguana at the headstart facility in Kingston, Jamaica. *Photos by Winty Marsden.*

The most common nematodes in captive, headstart, and free-ranging rock iguanas are pinworms (oxyurids) (Figure 6.5). They are generally non-pathogenic. However, the direct life cycle of this parasite leads to reinfection and heavy worm burdens in captive reptiles. Treatment of oxyurids with antihelminthics is controversial. Some parasitologists believe that oxyurids can play a beneficial role in breaking up dry vegetation in the hindgut of the iguana (E. Grenier, personal communication). Oxyurids are not susceptible to standard doses of antihelminthics such as fenbendazole, pyrantal, and ivermectin. Successful treatment requires high doses of drugs given repeatedly, which raises the threat of doing more harm than good when trying to treat pinworms. This dilemma can be avoided by reducing the rate of superinfection through good enclosure hygiene practices, periodic substrate replacement, and daily removal of fecal material.

Protozoa, including coccidia and *Entamoeba*, can be responsible for significant morbidity, including poor weight gains or weight loss, enteritis, and malabsorption. Coccidia can be a frustrating problem in reptile hatchlings housed at high density, although this problem has not been specifically identified in rock iguanas. Good husbandry practices are the first line of defense. Clinical cases can be treated with sulfa-based coccidiastats such as sulfadimethoxine. Such treatment will reduce the load of coccidia and improve the health of the patient, but is unlikely to eliminate the infection. The drug ponazuril has shown potential to clear coccidial infections in mammalian species, and shows promise in preliminary studies in lizards (Bogoslavsky, 2007). Good husbandry is the best defense against this disease problem.

Entamoebiasis is not a common disease in lizards, but when it occurs it is often fatal. Signs include anorexia, dehydration, and wasting. Bloody diarrhea can be seen, but animals are often found dead before this clinical sign is manifested. Lizards are at increased risk of infection when they are housed in close proximity to snakes and chelonians or when transferred between collections. Pre-mortem diagnosis can be difficult. Finding the amoeba on fecal cytology can be helpful, but false negatives are common. Diagnosis can be made from histopathology of a necropsy specimen. If amoebiasis is suspected, treatment should be initiated as early as possible. Metronidazole, iodiquinol, and paromomycin have all been used successfully, and a combination therapy is ideal. Systemic antibiotics to treat concurrent septicemia are often required. When a diagnosis

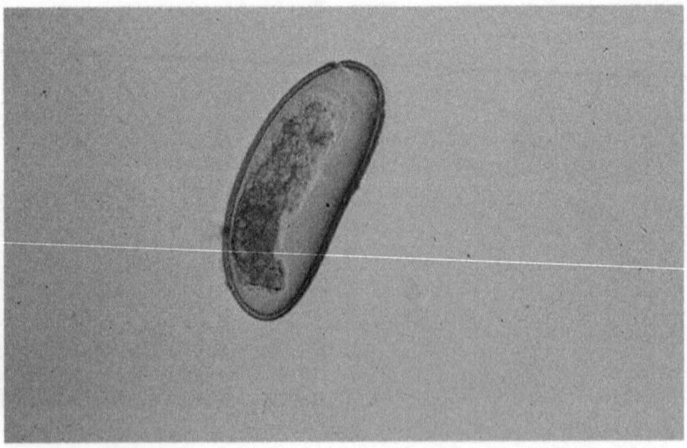

FIGURE 6.5 Oxyurid ova from *Cyclura* fecal flotation. 190 × 171 μm; magnification 160×. *Photo by Ellis Grenier.*

is made by histopathology, special attention should be given to other animals in the group.

Cryptosporidium has rarely, if ever, been found in captive or free-ranging rock iguanas and its capacity to cause disease in this group of lizards is not known.

Infections

The third most common reason for veterinary intervention in rock iguanas is infectious disease. These are predominately bacterial infections involving abscesses, pneumonia, and septicemia. Viral infections are rarely diagnosed in lizards. Fungal infections can be seen, but are much less common than invasion with opportunistic pathogenic bacteria.

Although the incidence of bacterial infection is very high, this is misleading. Animals kept in optimal housing and nutritional conditions rarely develop primary bacterial infections. However, other primary diseases such as those affecting the liver, heart, or genitourinary system, often go undetected, leading to debilitation and reduced immune function. The bacteria that ultimately kill the iguana are usually opportunistic invaders. A clinical or post-mortem evaluation that detects a bacterial infection should be backed up with a thorough evaluation of underlying pathology.

The majority of bacterial pathogens in iguanas are gram-negative rods such as *Pseudomonas* and *Salmonella*, although numerous bacterial species have been cultured from diseased iguanas, including *Proteus*, *Neisseria*, *Serratia*, *E. coli*, and others. Broad spectrum antibiotic coverage should include drugs that are effective against *Pseudomonas*, such as enrofloxacin (Baytril), aminoglycosides (amikacin or gentamycin), and third generation cephalosporins such as ceftazidime (Fortaz). When a *Pseudomonas* infection is confirmed by culture, antibiotic sensitivity testing must be done since drug resistance in this bacterium is common. Combination therapy using two of the three drugs listed will be more effective against *Pseudomonas* than any of the drugs alone. When antibiotic therapy is initiated prior to bacterial culture results, broad spectrum coverage that includes gram-negative, positive, and anaerobic bacteria is appropriate.

REPRODUCTIVE PROBLEMS

Reproductive disease in captive rock iguanas is common, and represents some of the most life-threatening conditions seen. The husbandry and nutrition of captive rock iguanas has improved in recent years as more has been learned about the normal biology of this group of animals. However, nutrition and husbandry-related challenges still contribute significantly to the high incidence of reproductive complications and there is still more to learn.

Rock iguanas are oviparous, producing a single clutch of pliable eggs each year. In the wild, folliculogenesis usually begins in March for most species, followed by nesting behavior in late May to early June, and egg laying in June to early July. These times may vary in more northern captive settings. Animals undergoing folliculogenesis should have higher than normal circulating calcium, phosphorus, cholesterol, and triglyceride levels. The presence of these biochemical variations is a reliable indicator that a female is in an active reproductive

TABLE 6.2 A Comparison of Calcium, Phosphorus, Cholesterol, and Triglyceride Levels in Reproductive and Non-reproductive Female Rock Iguanas

	Calcium (mg/dl)	Phosphorus (mg/dl)	Cholesterol (mg/dl)	Triglycerides (mg/dl)
Non-reproductive	12	6	85	115
Undergoing folliculogenesis	Can be >150	Can be >16	Can be >280	Can be >700

state. It is important to interpret plasma chemistry values within the framework of the reproductive cycle of the female patient. For example, normal, non-reproductive plasma calcium in a healthy iguana should be approximately 12 mg/dl, but can exceed 150 mg/dl in a reproductively active female (Table 6.2).

It is common for female iguanas to stop feeding during the weeks leading up to oviposition, to lose condition, and to look "spent" after a clutch has been laid. Throughout this period of anorexia, their attitude and nesting behavior should remain normal. A healthy female will rebound quickly and replace lost body condition. An anorexic female that becomes lethargic or fails to lay her eggs in a normal window of time should receive medical intervention.

Reproductive complications reported in rock iguanas include infections of the ovary and oviduct (oophoritis and salpingitis), failure to ovulate (follicular stasis, follicle binding), egg yolk peritonitis (coelomitis), failure to pass eggs (egg binding or dystocia), torsion of the oviduct, hemipene and oviductal prolapses, and neoplasia. All of these conditions are life-threatening and require skilled medical intervention (Divers, 2000; Jacobson, 2003; Mader, 2006; Juan-Salles et al., 2008).

Diagnostic evaluation should include a complete physical exam, review of reproductive, husbandry, and weight history, complete blood count and plasma chemistry profile, coelomic ultrasound, and whole body radiographs. Care should be taken when performing a diagnostic evaluation on a female undergoing folliculogenesis, as manual restraint creates a risk of follicle rupture leading to yolk peritonitis. A safe technique for handling involves luring the animal into a tube such as a PVC pipe, closing the ends, and administering isoflurane anesthesia into the tube. The iguana will relax and can be handled with less struggling. Ultrasound is used to evaluate the status of the ovary and to monitor the process of folliculogenesis. Once the follicles have ovulated and been shelled, radiographs can be used to confirm the presence and number of eggs and to determine if there is any physical obstruction preventing egg passage (Figure 6.6).

Oophoritis and salpingitis can be difficult to diagnose pre-mortem. Suspicion is high if the animal is reproductively active and if clinical and laboratory data support sepsis. Laparoscopic surgery can provide visual and microbiological support for the diagnosis. Treatment should include broad spectrum antibiotic coverage. Combination drug therapy is preferred. Husbandry and temperature conditions should be optimized. Fluid and nutritional support should be provided.

Follicular stasis is a common problem in green iguanas and has been seen in captive rock iguanas. In this condition, the follicles develop normally but fail to ovulate. The mechanism for this is unknown, but likely relates to subtle environmental or nutritional issues that are not optimized in the captive setting. Diagnosis is made by ultrasound exam, documenting

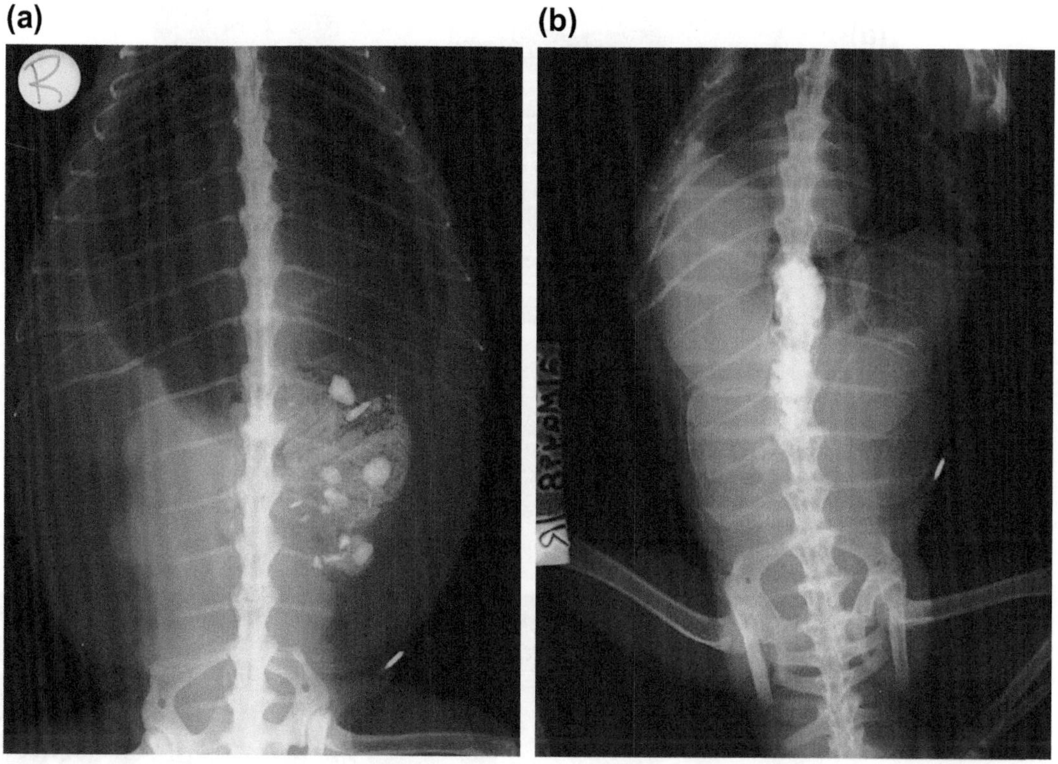

FIGURE 6.6 (*a*) Radiograph of a female Jamaican iguana with follicles. (*b*) Radiograph of gravid female Jamaican iguana with infertile, mis-shapen eggs.

a follicular phase that began at a normal time, but has progressed past when normal ovulation should have occurred. Radiographs can be used to confirm that there are no shelled eggs present. Hormone therapy to stimulate ovulation has not proven useful. Occasionally reptiles will resorb the yolk of unovulated follicles. However, the presence of the retained yolk predisposes the animal to ovarian infection, yolk emboli, and yolk coelomitis, all of which have a poor prognosis for recovery. Surgical intervention to remove the ovaries is the best life-saving option and must be considered even though it removes that animal from the breeding population (Backeus and Ramsay, 1994; Scheelings, 2008). Care must be taken to avoid accidental damage to the adrenal gland and local blood vessels during surgery (Figure 6.7).

Dystocia is a condition in which females fail to pass eggs at the appropriate time. Dystocia can be difficult to differentiate from normal pregnancy since anorexia and weight loss are part of a normal egg-laying period. Knowledge of the iguana's normal biology is critical. Sometimes a female is actively pushing and no eggs are being passed. These cases are easy to diagnose. More typically, a female will show weak to moderate nesting behavior over a period of time, yet fail to lay eggs. The first intervention is to make sure that the nesting environment and temperature are ideal. Many dystocias occur

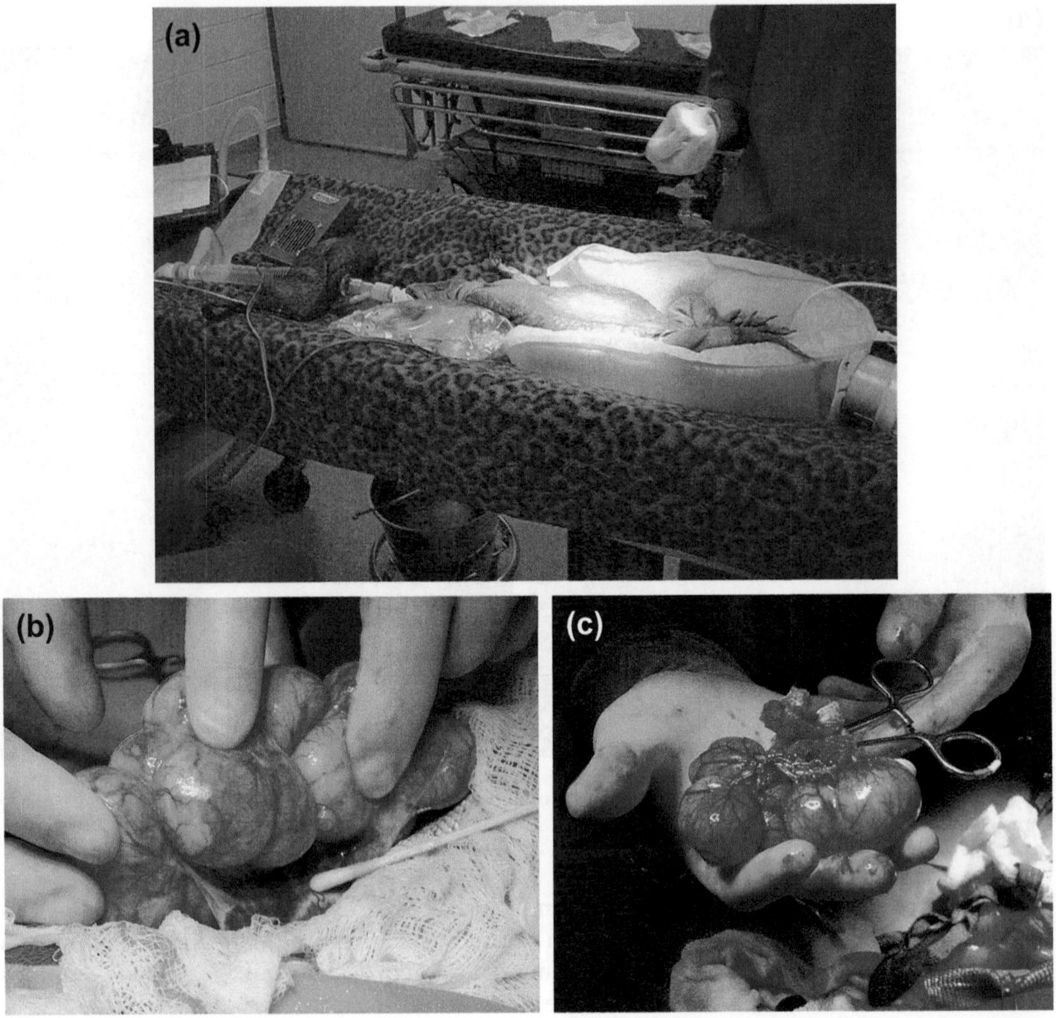

FIGURE 6.7 Ovariectomy in a Jamaican iguana with chronic follicular stasis. (*a*) The anesthesia and surgical set-up for a ventral paramedian approach. (*b*) The follicle-rich ovary being gently exteriorized, exposing the close proximity of the ovarian stump to the adrenal gland. (*c*) The ovary in hand after surgical removal. *Photos by Shannon Ferrell.*

because the female does not have all of the environmental conditions necessary to trigger the normal laying process. When medical intervention is elected, it is important to evaluate the whole patient and rule out concurrent disease or physical obstruction, correcting those problems as necessary. When the animal is in good condition and well hydrated, dystocia can be treated using oxytocin (published doses range from 1–30 IU/kg body weight). Best results are achieved when a second dose is given 45–90 minutes after the first. A percentage of dystocias that are refractory to oxytocin therapy can be reduced using arginine vasoticin, a more specific reproductive hormone in reptiles. This drug is experimental and expensive, but should be considered if oxytocin fails (Lloyd, 1990; Gross et al., 1992; Carpenter, 2005; Plumb, 2008).

When hormone therapy fails to correct a dystocia, surgical intervention is a good option. If the reproductive capability of the female can be sacrificed, then removing the ovaries is the most efficient surgical approach and will eliminate future reproductive complications. If the female's reproductive status needs to be preserved, then multiple small incisions in the oviducts will be necessary to remove the numerous eggs. If the animal is otherwise healthy and the reproductive tract is free of infection, the iguana should retain normal reproductive capability.

Yolk coelomitis occurs when yolk material leaks from a follicle and becomes free in the coelom. The animal's body treats the yolk as foreign material, generating a significant inflammatory response. Secondary bacterial coelomitis is common, followed by sepsis. The prognosis in these cases is poor. Animals that survive have a poor reproductive future. Treatment should include supportive care appropriate for sepsis, antibiotic and anti-inflammatory therapy, and possible surgical flushing of the coelom.

Hemipene prolapse generally occurs following breeding activity in the male iguana. In the female, oviductal prolapse can occur during egg laying. The principles of prolapse management are similar for both cases, with earlier intervention leading to better outcomes. The tissue needs to be kept moist, clean, and protected until it can be replaced to its normal position. General anesthesia is usually required to facilitate adequate relaxation. With gentle tissue handling and patience prolapses can usually be reduced. Appropriate antibiotic coverage should be provided and the animal should be observed for recurrence. If the prolapsed tissue is necrotic, surgical removal will be necessary.

Preventive medicine is the key to maintaining healthy reproductive tracts in reptiles. Future research focusing on the normal biology of these animals will enable further improvement of husbandry practices. It is important to monitor vitamin D and calcium status closely, maintain optimal temperature ranges and photoperiods, and provide ideal nesting environments in order to avoid continued reproductive problems in captive rock iguanas.

Renal Disease

Iguanas have paired kidneys that are located deep in the pelvic canal. This anatomic location makes the normal iguana kidney difficult to evaluate by palpation, radiography, or ultrasound. Renal disease is common in iguanas and can result in frustrating clinical cases. Chronic renal disease may be suspected in patients that show slow, chronic demise, poor body condition, anorexia, and lethargy. Acute renal failure may be suspected in an animal with excellent, robust state with acute lethargy, anorexia, reduced urine output, and elevated plasma uric acid.

Chronic renal disease may occur secondary to high protein diets, inadequate humidity resulting in chronic subclinical dehydration, hypervitaminosis D, and chronic infections. Acute renal failure can occur from a toxic insult such as nephrotoxic plants or improper use of nephrotoxic drugs such as amikacin and gentamycin. A bacterial infection of the kidney has the potential to cause acute renal disease. Kidneys that are not functioning properly will fail to clear uric acid from the blood, resulting in the accumulation of urates in and on the kidneys and on the surface of other organs (visceral gout).

Complete blood count and plasma chemistry analysis should be performed when evaluating an iguana with suspected renal disease. However, there are limitations to the utility of blood results in these cases. Creatinine and BUN are not helpful in evaluating renal function because the nitrogenous waste product of the iguana is uric acid, not urea. In acute renal failure, the uric acid can be quite elevated, resulting in acute visceral gout due to precipitation of the uric acid crystals on organ surfaces. In chronic renal failure, the plasma uric acid is often normal to slightly elevated. However, other changes such as anemia and elevated phosphorus can raise the suspicion of renal disease. Evaluation of plasma electrolytes is of limited value in cases of renal disease because iguanas regulate electrolytes through a number of organs, including the kidneys, nasal salt gland, cloaca, colon, and urinary bladder. The best diagnostic tool for evaluating renal health in the iguana is a renal biopsy. Several techniques have been described for safely accessing the intrapelvic iguana kidney (Divers, 2000; Divers and Innis, 2006).

Calcium and Vitamin D Related Problems

Maintenance of calcium and vitamin D homeostasis in captive rock iguanas has been difficult in many species across institutions. Despite improvements in husbandry and nutrition, this problem is stubbornly persistent and likely contributes to a number of secondary health problems in captive rock iguanas, including poor reproductive success, metabolic bone disease, orthopedic fractures, and hypocalcemic tetany (Laing et al., 2001; Bernard et al., 2006; Mader, 2006). Although some anecdotal information exists, a broad review of husbandry practices at holding institutions is needed to better understand the factors that result in hypocalcaemia and hypovitaminosis D (Bernard et al., 2006).

Rock iguanas in captivity generally have lower plasma vitamin D3 levels than their wild and headstart counterparts (N. Lung, personal observation). This likely results from inadequate dietary sources of vitamin D, inadequate exposure to natural ultraviolet light in indoor enclosures, and inadequate wavelength, distance, and duration of exposure to artificial ultraviolet light (Allen et al., 1999; Laing and Fraser, 1999; Laing et al., 2001; Ferguson et al., 2009). Even in completely outdoor enclosures, such as those in many headstart facilities, plasma vitamin D levels can decline when natural vegetation shades the enclosure, reducing direct sunlight (N. Lung, unpublished data). It is important to keep trees and vines trimmed in order to ensure direct sunlight to part of each outdoor area, and to be cognizant of competition for basking spots in group housing as some individuals can be out-competed for access to direct sunlight.

Intestinal Impaction

Intestinal impaction in rock iguanas has been recorded in at least ten animals. Impaction can occur after ingestion of a large foreign body. However, impaction is more commonly the result of sand or grit ingestion during feeding and foraging (Figure 6.8). Sand impactions will usually present as anorexia and lethargy with reduced fecal output. Straining to defecate is not always present. Intestinal impaction can also result secondary to colonic compression

FIGURE 6.8 Placing food dishes above the substrate on trays will help minimize intake of sand and dirt, which should minimize the risk of intestinal impactions.

from a dystocia or urinary calculus. Radiographs should be used to evaluate the underlying cause of an impaction and treatment initiated accordingly.

Cystic Calculi

Iguanas have a fully developed urinary bladder that is fed by a ureter from each kidney, and drained by a single urethra that empties at the urodeum in the cloaca. As is true for many reptiles, bladder stones, or cystic calculi, are a common medical problem in iguanas (Figure 6.9). They have been seen in rock iguanas in zoos, as well as in headstarting facilities. The etiology is poorly understood but is likely multifactorial including nutrition (vitamin, mineral, and moisture content of the diet), humidity and chronic low grade dehydration, and possibly urinary tract infections. When a cystic calculus is identified, it should be surgically removed. Untreated stones can lead to trauma to the bladder wall, urinary infections, urinary obstruction, egg binding, and fecal retention.

Diagnosis is made through coelomic palpation and radiography. Larger stones are easily palpated in the caudal coelom. Urate stones can be radiolucent. However, the chemical makeup of these stones is quite variable and there is usually enough mineral content in the stone to make it radiodense.

Surgical removal of the stone involves a straightforward cystotomy using a ventral paramedian approach – similar to that used for ovariectomy but somewhat more caudal (Figure 6.10). Chemical analysis of the stone is recommended to add to current knowledge of this disease process. With maintenance of adequate hydration and broad spectrum antibiotic coverage, recovery should be uneventful (Mader, 2006).

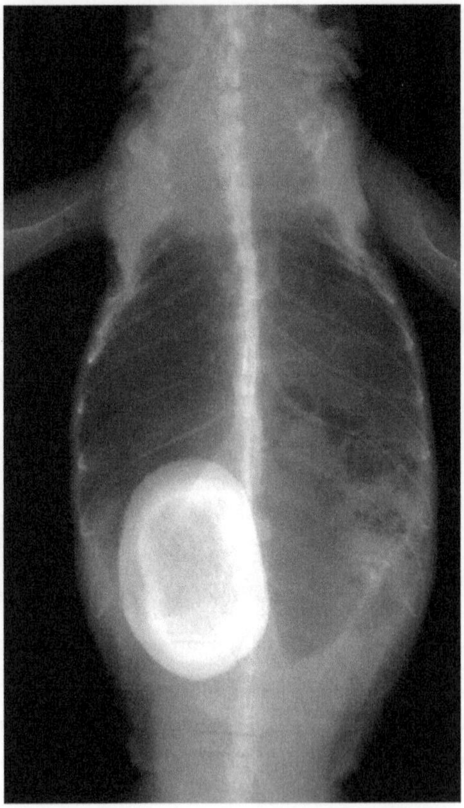

FIGURE 6.9 Radiograph of a Jamaican iguana with a bladder stone. *Photo by Nancy Lung.*

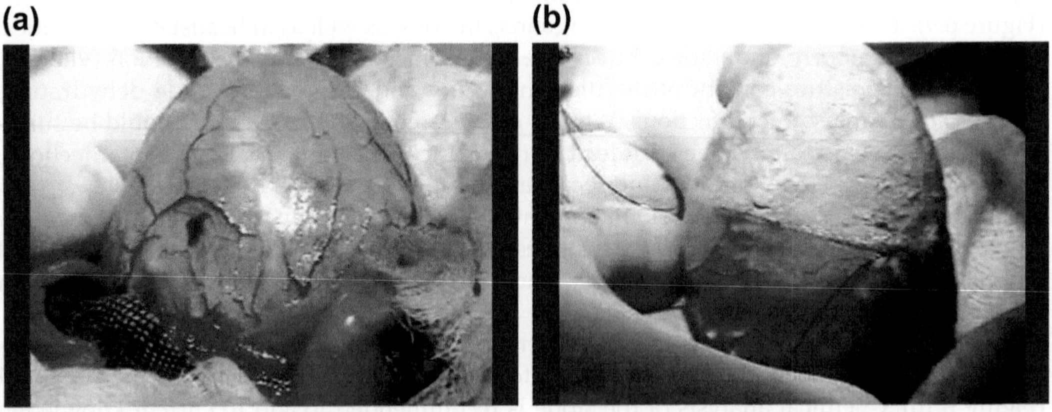

FIGURE 6.10 Surgery to remove the stone is shown – first with the bladder intact (*a*), second with the bladder incised and the stone being gently extracted (*b*). *Photos by Shannon Ferrell.*

Idiopathic Leukocytosis

In recent years, one zoological institution has dealt with an unusual leukocytosis (elevated white blood cell count) in their Jamaican iguana population. Both male and female iguanas were affected. White blood cell counts were as high as 72,000/ml, detected during a routine exam. The animals were clinically normal. White blood cell counts subsequently returned to normal with antibiotic therapy, but some animals returned to normal without any treatment. There is not an apparent seasonality to the high cell counts. The Grand Cayman Blue and Rhinoceros iguanas housed at the same facility under similar husbandry conditions were unaffected. The cause of the periodic leukocytosis remains a mystery (J. Ramer and J. Proudfoot, personal communication).

CLINICAL TECHNIQUES

Blood Collection

A reliable measure for how much blood can safely be collected from a lizard is to use an upper limit of 0.7–1% of the animal's body weight. For example, from a 100 g animal, 0.7–1 g of blood, which is equivalent to 0.7–1 ml of blood, can safely be collected. This calculation is important when working with hatchling and juvenile iguanas, since many laboratory blood tests require a minimum volume of blood for analysis. While some laboratories have equipment for working with small sample volumes, others do not, and only a small number of laboratories are skilled with reptilian blood. A high quality laboratory should be capable of running a full complete blood count and plasma chemistry panel from 0.5 ml blood. For most laboratory analysis (complete blood count, blood chemistry, nutritional analysis such as vitamin D) a lithium heparin blood tube should be used. The lithium heparin will keep the blood sample from clotting and will not interfere with any of the laboratory assays.

The most reliable site from which to collect blood from iguanas is the ventral coccygeal vein. This vein runs on the midline of the tail just ventral to the vertebral bodies. To avoid potential damage to the hemipenes, this vein should be accessed a few inches distal to the cloaca. Both the lateral and ventral approaches are useful, although the lateral approach seems to be more successful in rock iguanas.

For the lateral approach, the animal should be in a normal resting position on a flat surface with a handler controlling the head and pelvic region (Figure 6.11). The sides of the tail can be palpated to identify the lateral aspect of the vertebral transverse processes (sometimes this can be seen as a small indentation on the tissue on the lateral side of the tail). The area should be cleaned with alcohol to remove dirt and debris. The needle should then be inserted two to three scales ventral to the lateral indentation and inserted perpendicular to the skin to a depth half the width of the tail. If the needle reaches bone, it should be redirected slightly lower. If the needle goes past midline and has not penetrated the vein, it should be redirected slightly higher.

For the ventral approach, the animal should be held upright or on its back with the ventral side facing the collector, and the tail should be held to stabilize it (Figure 6.12). Alcohol should be used to clean a region a few inches distal to the cloaca. The needle should be inserted perpendicular to the vertebral column exactly on the midline, and advanced using slight

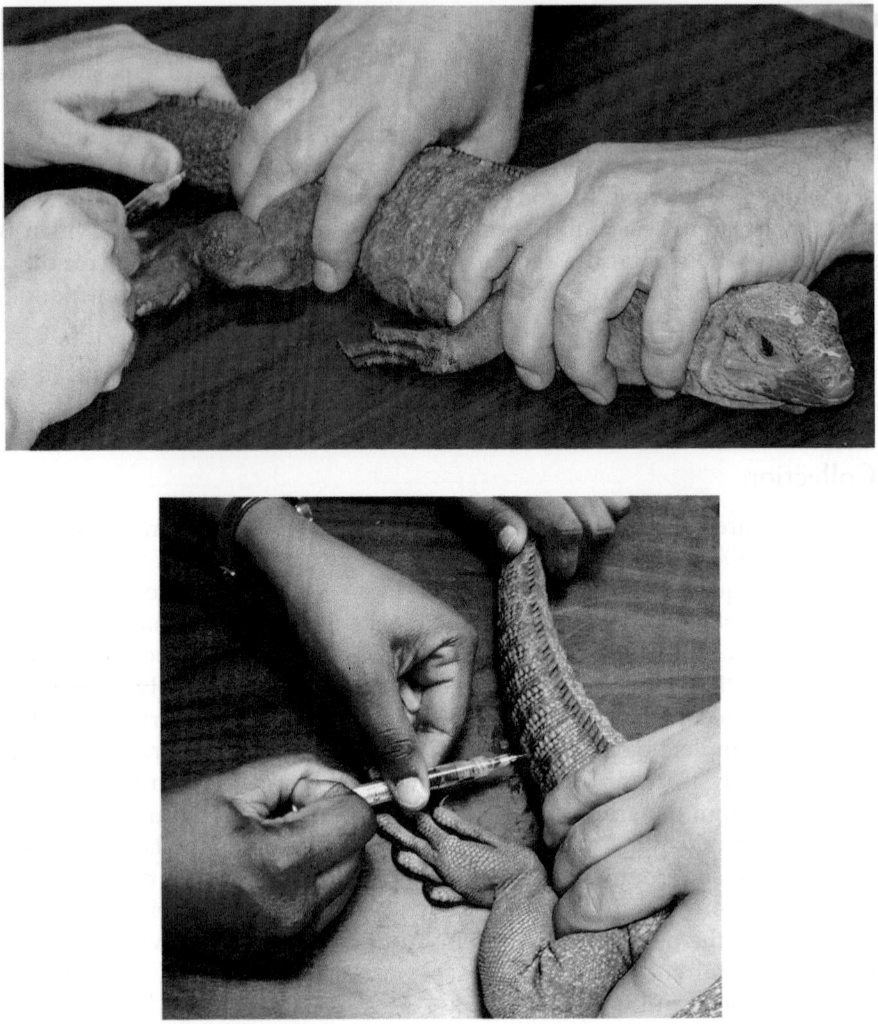

FIGURE 6.11 Lateral approach to blood draw from the ventral tail vein. Note the restraint and the positioning of the needle. *Photos by Nancy Lung.*

negative pressure on the plunger until blood starts to flow or until the needle reaches bone. If the needle reaches bone, it should be backed out slightly and/or redirected onto the midline.

Gavage Feeding

Gavage feeding, or "stomach tubing," is a key method of providing support to an ill iguana (Figure 6.13). With experience, it can be used repeatedly with low stress to the animal and will improve the outcome of many difficult cases. Tubing is routinely used to provide oral medication to reluctant or anorexic animals, and to provide nutritional and fluid support

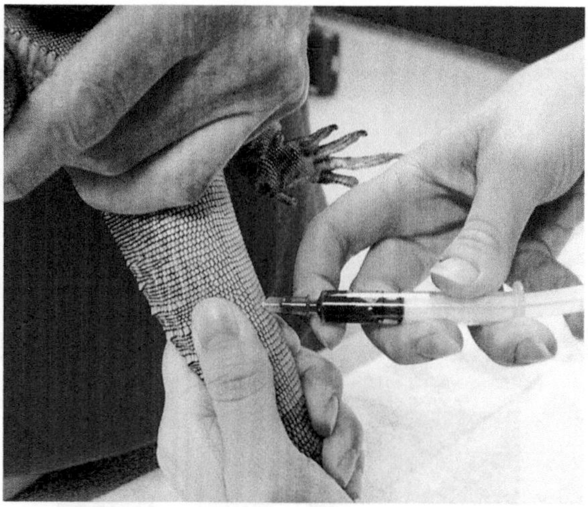

FIGURE 6.12 Ventral midline approach to blood collection. *Photo by Nancy Lung.*

to an ill or anorexic patient. The method can also be used as a diagnostic tool for collecting samples of gastric contents.

Both flexible and rigid gavage tubes can be used successfully (Figure 6.14). Before starting the procedure, estimate the distance from the mouth to the stomach to provide a target for the distance the tube will need to travel in order to reach the stomach. If the tube is not inserted far enough, the infused solution is likely to reflux into the oral cavity, risking aspiration. If the tube is inserted too far, damage or perforation of the stomach wall can occur.

Most iguanas will spontaneously open their mouths defensively during manual restraint. This opportunity can be taken to insert a soft-coated speculum. Once an iguana closes its

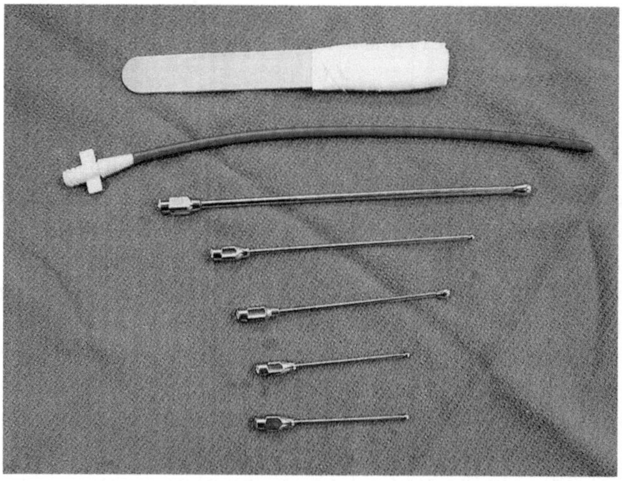

FIGURE 6.13 Gavage feeding tools. *Photo by Nancy Lung.*

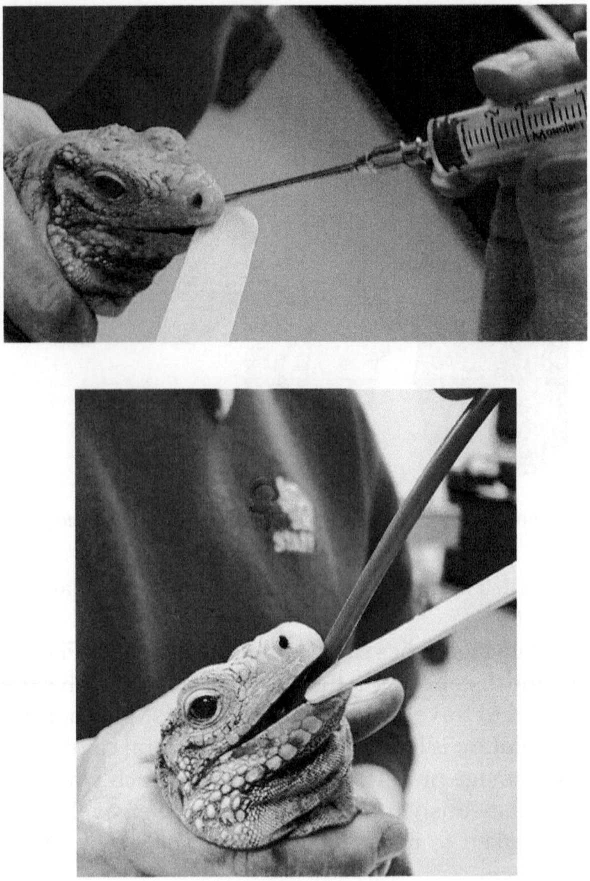

FIGURE 6.14 Both rigid and flexible tubes can be used in iguanas. A soft mouth speculum should be used to protect the teeth from being damaged when the animal bites down on the feeding tube. *Photos by Nancy Lung.*

mouth tightly it is difficult to pry it open without risking injury to the animal or the handler, although gentle downward pressure on the dewlap can be helpful. The soft coating on the speculum helps to protect the animal's teeth. The feeding tube can be inserted into the pharynx above or to the side of the glottis, which can be visualized just behind the tongue. The tube is then advanced to the desired depth, and the product infused slowly. While infusing, it is important to watch for evidence of reflux. If reflux occurs, infusion should be discontinued and the tube removed. Aspiration is more likely when an animal is struggling during manual restraint. If the animal is set down onto a flat surface with a mouth full of feeding material, it will relax and swallow normally or shake the material out of its mouth. When gavage feeding is one of several treatments during a manual restraint, the gavage treatments should be saved for last in order to reduce the incidence of regurgitation.

When using tube feeding in the medical management of anorexic, ill iguanas, a safe rule of thumb is to feed no more than 1.5–2% of the animal's body weight at a time. This equates to

15–20 g of formula per kilogram of body weight. Commercially available products that give good clinical results include Ensure Plus (Ross Laboratories, Columbus, OH), and Oxbow critical care for herbivores (Oxbow Hay Company, Murdock, NE). A reasonable short-term home-made enteral formula consists of 99.75% vegetable baby food and 0.25% ground adult multivitamin and mineral tablets such as Centrum. The baby foods with the best calcium to phosphorus ratio include green beans, carrots, squash, and garden vegetable. Baby foods to avoid include peas, mixed vegetable, spinach, sweet potato, fruits, and any meat-based baby food. The energy density and consistency of this formula can be improved by adding ground high-fiber monkey chow (Donoghue, 2006).

The calculation of how much formula is required comes from a general reptile equation for energy needs: kcal/day for standard metabolism $= 32 \times (\text{body weight in kg})^{0.75}$. For Enteral Herbivore, this equates to 1.4 kcal per g of formula. For example, for a 2 kg iguana: kcals to feed for baseline metabolism $= 32 \times 2^{0.75} = 32 \times 1.68 = 53.76$ kcal; $53.76\text{ kcal} \div 1.4\text{ kcal/g} = 38.4$ g of formula per day. This is a manageable volume and may be divided into two feeds per day if needed (A. Ward, personal communication).

Cloacal Swabbing

A cloacal swab can be used to collect a sample for bacterial culture, or to obtain fresh fecal material for cytology (Figure 6.15). The iguana can hold the cloaca tightly shut, so swabs must be inserted gently, ensuring not to push too hard and risk injury. One hand can be used to gently spread the opening to the cloaca, while the other hand is used to insert a cotton-tipped applicator directly perpendicular to the tail for initial access. Sometimes it is necessary to direct slightly caudally into the vent to gain initial access. Once the tip has passed into the opening, it should be re-directed in a forward direction to a depth of about an inch. Finally, the cotton tip should be rolled back and forth to ensure good mucosal contact, and then slowly withdrawn.

Euthanasia

Euthanasia may be required in the management of captive rock iguanas for a number of reasons, including alleviating suffering in critically ill animals and culling non-viable hatchlings. Numerous techniques have been employed in the euthanasia of reptiles. However, many are inhumane and/or detrimental to good post-mortem evaluation. Physical methods of euthanasia include deep freezing, decapitation, and pithing through the parietal eye. All physical methods of euthanasia must be preceded by a surgical plane of anesthesia, which can be achieved through excessive doses of injectable anesthetics (telazol or ketamine) or through deep inhalant anesthesia using isoflurane. Freezing is not recommended, as it results in a specimen that is of little use for post-mortem histopathology evaluation.

The ideal method for euthanasia is chemical, using either potassium chloride or a barbiturate solution. These should be administered intravenously or intracoelomically after a surgical plane of anesthesia has been reached as described above. For more information, see the guidelines from the American Veterinary Medical Association Panel on Euthanasia (http://www.avma.org/issues/animal_welfare/euthanasia.pdf).

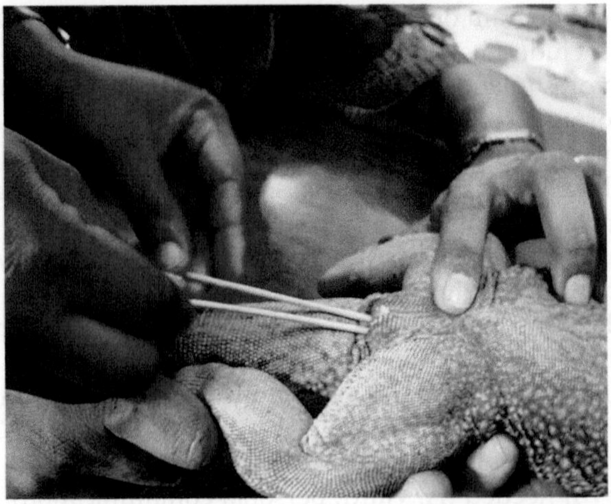

FIGURE 6.15 Passing sterile swabs into the cloaca of a rock iguana for the collection of material for bacterial culture. *Photo by Nancy Lung.*

HEALTH OF FREE-RANGING AND HEADSTART ROCK IGUANAS

Comprehensive health evaluations have been conducted on most species of West Indian rock iguanas. A database has been established that includes normal blood values, internal and external parasites, normal gastrointestinal bacterial flora, common health problems, growth rates, and nutrition. Such data are useful in the management of conservation programs and provide reference information for the medical management of iguanas in headstarting facilities, zoos, and private collections.

Health assessments of iguanas in the field include free-ranging iguanas as well as animals housed in headstarting facilities in range countries (Figure 6.16). Typically, such assessments include: (1) physical exams, (2) standard weight and body size measurements, (3) collection of whole blood for genetic analysis, white blood cell counts and differentials, screening for blood parasites, and measurement of hematocrit and total solids, (4) collection of plasma for measurement of biochemical parameters, (5) cloacal swabs for bacterial culture to assess normal intestinal bacterial flora and to screen for bacterial pathogens, and (6) collection of feces for assessment of intestinal parasite burdens.

Health assessments of iguanas in range countries pose unique challenges for each species. On the islands of Jamaica and Grand Cayman, there is good infrastructure and support for the iguana headstarting facilities, but access to free-ranging animals has been infrequent and unpredictable. This results in a database that is rich in data from headstarted animals, but poor in data from free-ranging animals. When wild iguanas are caught in the Hellshire Hills of Jamaica, it is generally during the nesting season, skewing the data toward reproductively active females (N. Lung, personal observation). For the San Salvador, Allen Cays, and Ricord's iguanas, all health assessments were performed on free-ranging animals (James et al., 2006; Maria et al., 2007; S. Murray and N. Lung, unpublished data). Although similar

FIGURE 6.16 Dr. Bonnie Raphael processes medical samples at the headstart facility in Grand Cayman. *Photo by John Binns, IRCF. org*

challenges exist in the evaluation of the other rock iguana species, the existing database is still an exceptional tool in the understanding of health management in these animals.

In general, the health of free-ranging rock iguanas is excellent. Exceptions to this occur in populations that live in marginal habitat. This may be due to competition for food resources with domestic animals or loss of normal forage due to the introduction of non-native vegetation. In these cases, body condition of the animals is less than ideal. Blood values will typically show chronic anemia, with packed cell volumes of less than 25%.

A unique condition appears to exist in the Mona Island iguana population, where problems with cataracts and other ophthalmic conditions occur in free-ranging adults (T. Reichard, personal communication). One theory for the high incidence of cataracts in this population is a normal age-related process in a population that is skewed toward older animals due to heavy predation on hatchlings. However, similar skewed age structures occur in other species of rock iguanas and cataracts have not been identified in these populations. Nutritional, genetic, and other factors should be investigated.

In general, the health of the rock iguanas in headstarting facilities is quite good based on data gathered over years of health screening. However, husbandry and nutrition issues present constant challenges. Daily diligence is needed in all programs to ensure that animals are housed at an ideal density and in proper age and sex ratios, observed for signs of illness, and fed a fresh, balanced, and complete diet. Although these concepts seem straightforward, they can be difficult to ensure at headstarting facilities in remote locations. It is important to perform annual health evaluations, including review of husbandry and nutrition practices. Strong relationships with range country veterinarians can help ensure that ill or injured animals receive timely medical care. Despite the challenges, the benefits of headstarting iguanas in their range country, including exposing them to the climate they will face in the wild and feeding them native vegetation, are significant.

Establishment of Normal Blood Values

Collection of blood samples (Figure 6.17) from headstarted and free-ranging rock iguanas has achieved two important goals. First, it has allowed veterinarians to evaluate the candidacy of headstarted individuals for release, i.e., determining if animals are in good health prior to release. Second, it has generated important data on normal blood values for several rock iguana species (Table 6.3). This database provides a tool for evaluating the health of individuals and populations, and is a useful reference for the medical management of iguanas in zoo and private collections.

Blood work is an integral part of pre-release health screening for headstarted iguanas. Because iguanas in marginal health tend to show anemia before other changes are apparent, the best internal measure of overall health in the iguana seems to be packed cell volume. It is not uncommon for an iguana in poor body condition with a packed cell volume of 25% or lower, to exhibit a normal plasma chemistry. When anemia is present, the animal should not be approved for release and further investigation into underlying health problems should be pursued. Blood results should be interpreted as in any clinical case, considering indicators for renal and liver status, occult infection, etc. Calcium, phosphorus, cholesterol, and triglycerides should be interpreted in light of sex and reproductive status.

Bacterial Flora of the Intestinal Tract

Cloacal swabs for bacterial culture should be part any health assessment program. Several years of data collection have resulted in an understanding of normal bacterial flora in the

FIGURE 6.17 Dr. Glenn Gerber (San Diego Zoo Global) and Jeff Lemm take a blood sample from a Turks and Caicos iguana.

TABLE 6.3 Normal Blood Values for Eight Taxa of Rock Iguanas

Parameter	*C. collei*	*C. pinguis*	*C. lewisi*	*C. ricordii*[1]	*C. c. stejnegeri*	*C. r. rileyi*	*C. n. nubila*[2]	*C. c. inornata*[3]
WBC (×1000/ml)	—	—	4.01 ± 2.53 (n = 32)	6.56 ± 1.84 (n = 19)	—	—	7.83 ± 1.53 (n = 13)	6.91 ± 2.57 (n = 18)
Heterophils (%)	57.2 ± 10.7 (n = 65)	61.9 ± 15.7 (n = 62)	58 ± 13 (n = 32)	76.1 (n = 16)	55.2 ± 24.7 (n = 19)	—	49.46 ± 6.81 (n = 13)	4.25 ± 1.84 (n = 18)
Lymphocytes (%)	8.1 ± 6.3 (n = 65)	14.8 ± 10.6 (n = 62)	16 ± 7 (n = 32)	7.37 (n = 16)	6.71 ± 3.7 (n = 19)	—	35.15 ± 7.67 (n = 13)	1.28 ± 0.68 (n = 18)
Azurophils (%)	24.6 ± 9.1 (n = 65)	17.6 ± 9.1 (n = 62)	14 ± 5 (n = 32)	7.1 (n = 16)	14.8 ± 5.1 (n = 19)	—	9.62 ± 1.37 (n = 13)	0.04 ± 0.06 (n = 18)
Eosinophils (%)	1.0 ± 1.9 (n = 65)	2.8 ± 3.1 (n = 62)	3 ± 2 (n = 32)	3.9 (n = 16)	5.9 ± 4.0 (n = 19)	—	0.46 ± 0.18 (n = 13)	0.12 ± 0.17 (n = 18)
Basophils (%)	8.2 ± 5.2 (n = 65)	1.6 ± 2.1 (n = 62)	9 ± 7 (n = 32)	2.1 (n = 16)	4.0 ± 3.3 (n = 19)	—	5.23 ± 0.80 (n = 13)	0.52 ± 0.30 (n = 18)
Monophils (%)	0.8 ± 1.5 (n = 65)	1.3 ± 2.7 (n = 62)	0 ± 0 (n = 32)	1.1 (n = 16)	0 ± 0 (n = 19)	—	0.00 ± 0.00 (n = 13)	0.24 ± 0.19 (n = 18)
PCV (%)	38.15 ± 5.27 (n = 75)	—	32.7 ± 6.6 (n = 24)	31.9 (n = 7)	—	34.6 ± 4.2 (n = 18)	29.12 ± 0.74 (n = 13)	29.06 ± 3.70 (n = 37)
AST (%)	31.42 ± 23.88 (n = 100)	23.81 ± 11.37 (n = 62)	69.6 ± 79 (n = 47)	39.7 ± 28 (n = 23)	28.43 ± 35.3 (n = 35)	31.8 ± 22.3 (n = 21)	—	29.47 ± 16.38 (n = 37)
CK (%)	1207 ± 1380 (n = 85)	2136 ± 1887 (n = 62)	—	2775 ± 264 (n = 23)	1536 ± 2149 (n = 33)	1085 ± 646 (n = 20)	3738.5 ± 504.3 (n = 16)	234 ± 2572 (n = 37)
Total protein (g/dl)	7.24 ± 1.29 (n = 100)	6.19 ± 1.17 (n = 62)	7.2 ± 1.4 (n = 47)	—	7.43 ± 1.07 (n = 35)	5.99 ± 1.0 (n = 21)	6.7 ± 0.4 (n = 16)	4.80 ± 0.88 (n = 37)
Albumin (g/dl)	3.20 ± 0.42 (n = 95)	2.69 ± 0.33 (n = 62)	2.8 ± 0.4 (n = 47)	2.13 ± 0.31 (n = 23)	2.27 ± 0.38 (n = 35)	3.18 ± 0.64 (n = 21)	2.6 ± 0.1 (n = 16)	2.03 ± 0.38 (n = 37)
Globulin (g/dl)	4.15 ± 1.10 (n = 90)	3.50 ± 0.95 (n = 62)	4.5 ± 1.3 (n = 47)	5.1 ± 0.89 (n = 23)	5.16 ± 0.84 (n = 35)	2.81 ± 0.52 (n = 21)	—	2.86 ± 0.59 (n = 37)
BUN (mg/dl)	0.44 ± 1.1 (n = 50)	0.18 ± 0.49 (n = 45)	1.1 ± 1.2 (n = 47)	1.3 ± 0.82 (n = 23)	1.20 ± 1.11 (n = 35)	0.29 ± 0.52 (n = 21)	2.1 ± 0.1 (n = 16)	—
Cholesterol (mg/dl)	95.96 ± 49.25 (n = 52)	61.6 ± 19.8 (n = 45)	90.1 ± 40.4 (n = 47)	55.43 ± 38.25 (n = 23)	118.9 ± 73.05 (n = 33)	122.9 ± 53.4 (n = 21)	82.1 ± 9.6 (n = 16)	96.81 ± 33.71 (n = 37)
Glucose (mg/dl)	201.1 ± 33.3 (n = 101)	215.6 ± 63.6 (n = 62)	195.7 ± 52.7 (n = 47)	222.6 ± 49.8 (n = 23)	175.2 ± 57.2 (n = 35)	173.9 ± 27.1 (n = 21)	229.9 ± 8.0 (n = 16)	189.2 ± 39.4 (n = 37)
Calcium (mg/dl)	11.59 ± 2.40 (n = 99)	11.64 ± 2.66 (n = 60)	17.0 ± 20.7 (n = 47)	12.79 ± 2.07 (n = 23)	13.23 ± 3.5 (n = 31)	12.3 ± 1.34 (n = 21)	13.7 ± 1.2 (n = 16)	10.06 ± 0.95 (n = 20)
Phosphorus (mg/dl)	6.01 ± 1.36 (n = 86)	6.56 ± 1.30 (n = 60)	8.1 ± 2.7 (n = 47)	5.55 ± 1.22 (n = 23)	4.79 ± 1.59 (n = 33)	4.84 ± 1.82 (n = 21)	5.8 ± 0.3 (n = 16)	4.72 ± 1.42 (n = 20)
Chloride (mEq/l)	113.9 ± 12.6 (n = 90)	127.7 ± 23.8 (n = 62)	114.0 ± 5.5 (n = 47)	102.4 ± 11.4 (n = 23)	106.5 ± 9.3 (n = 35)	122.4 ± 9.2 (n = 21)	125.3 ± 2.1 (n = 16)	118.5 ± 7.4 (n = 37)
Potassium (mEq/l)	2.63 ± 1.34 (n = 82)	2.57 ± 1.45 (n = 62)	3.0 ± 2.0 (n = 47)	3.52 ± 1.27 (n = 23)	2.83 ± 0.94 (n = 35)	2.18 ± 0.82 (n = 21)	4.1 ± 0.1 (n = 16)	3.66 ± 1.06 (n = 37)
Sodium (mEq/l)	170.5 ± 10.0 (n = 94)	166.8 ± 6.1 (n = 62)	175.5 ± 16.1 (n = 47)	162.8 ± 14.8 (n = 23)	161.88 ± 8.95 (n = 35)	176.29 ± 11.09 (n = 21)	165.4 ± 2.2 (n = 16)	166.92 ± 5.99 (n = 37)
Uric acid (mg/dl)	4.59 ± 2.06 (n = 101)	4.38 ± 1.78 (n = 62)	4.8 ± 1.9 (n = 47)	3.55 ± 2.05 (n = 23)	3.10 ± 1.46 (n = 35)	2.58 ± 2.28 (n = 21)	4.8 ± 0.5 (n = 16)	1.77 ± 2.0 (n = 37)
Triglycerides (mg/dl)	144.4 ± 154.2 (n = 76)	120.5 ± 97.1 (n = 62)	221.6 ± 222.5 (n = 47)	69.8 ± 70.0 (n = 23)	2.06.8 ± 244.7 (n = 32)	94.43 ± 126.98 (n = 21)	159.8 ± 40.5 (n = 16)	103.5 ± 75.45 (n = 20)
Bile acids	12.57 ± 13.47 (n = 38)	6.35 ± 3.9 (n = 45)	6.8 ± 5.1 (n = 47)	7.28 ± 7.03 (n = 23)	6.61 ± 6.66 (n = 35)	26.21 ± 16.11 (n = 21)	—	10.62 ± 20.24 (n = 37)
Vitamin D (25 OHD) (ng/ml)[4]	145 ± 58 (n = 34) (HS)	319 ± 95.6 (n = 61) (HS)	272.8 ± 134.8 (n = 29) (HS)	222.4 ± 113.0 (n = 22) (W)	142.4 ± 64.26 (n = 10) (HS)	162.8 ± 57.4 (n = 21) (W)	—	—
	157 ± 137 (n = 14) (W)	193.6 ± 35.4 (n = 10) (W)	215 ± 78 (n = 2) (W)					

Means and standard deviations are shown, with sample sizes in parentheses.
HS = headstarted; W = wild.
References: [1]Maria et al., 2007; [2]Alberts et al., 1998; [3]James et al., 2006; [4]Ramer et al., 2005.

digestive tracts of Jamaican, Anegada, Mona Island, Grand Cayman Blue, and Ricord's iguanas. In this longitudinal study, samples from each group of animals were processed at one laboratory using consistent methodology. Differences between species were identified, as well as differences across years within species. All groups showed a broad range of mixed bacterial flora of both gram-positive and gram-negative varieties. The most common isolates included: *Enterococcus* spp., *Escherichia coli*, *Bacillus* spp., *Enterobacter*, *Salmonella* (more than 20 serotypes), *Corynebacterium*, *Clostridium* spp., and *Staphylococcus* spp. *Pseudomonas aeruginosa*, a common pathogen in reptiles, was rarely identified in the any of the iguana species examined.

Salmonella can be pathogenic in many animals across taxa. In reptiles, salmonella can cause infection as an opportunistic pathogen, but is usually found as part of the normal gastrointestinal flora. There appears to be wide variation in positivity rates for salmonella in rock iguanas both across populations, as well as temporally within a species (James et al., 2006; Maria et al., 2007; N. Lung, unpublished data).

Parasites

Ticks and mites are commonly found on free-ranging rock iguanas, sometimes in high numbers (Figure 6.18). Headstarted iguanas tend to be cleaner than their wild counterparts, with only occasional ticks or mites identified.

Oxyurids, or pinworms, are a ubiquitous intestinal nematode found in most rock iguanas, particularly in free-ranging animals, and are considered to be commensal and non-pathogenic

FIGURE 6.18 Ticks, such as the one seen on the dewlap of this wild Cuban iguana, are common ectoparasites of wild iguanas.

(James et al., 2006). Consumption of feces is common among young lizards, facilitating the transmission of intestinal parasites. Headstarted iguanas are generally infected with oxyurids, although headstarted Cuban iguanas were free of intestinal parasites (Alberts et al., 1998).

Coccidia are occasionally found in rock iguana fecal samples, including the Mona Island iguana and the Cuban iguana. Although coccidia have the potential to cause illness in reptiles, this has not been a significant problem in rock iguanas to date. More work needs to be done to identify the species of coccidia that invade the intestinal tract and evaluate their pathogenic potential.

Additional work also needs to be done to evaluate the prevalence and pathogenicity of hemoparasites in free-ranging rock iguanas. Alberts et al. (1998) reported a high percentage of the erythrocytes of both headstart and free-ranging Cuban iguanas infected with the piroplasm *Sauroplasma*, most likely transmitted by arthropod vectors such as ticks or mosquitoes. While heavy hemogregarine infection can lead to anemia, light to moderate infections are usually not pathogenic in their natural hosts. Further work is needed to determine the range and clinical significance of hemoparasites in rock iguana populations.

Calcium and Vitamin D

Health problems related to calcium and vitamin D are uncommon in free-ranging and headstarted iguanas, likely reflecting the benefit of maintaining headstart programs in range countries, using diets that include native vegetation, in open air enclosures that permit adequate access to natural sunlight. In headstart facilities, it is important to keep trees and vegetation from over-growing the headstart enclosures to insure that animals have adequate access to prime basking spots.

Normal circulating plasma vitamin D levels are shown in Table 6.3. Note that the reported values are for 25 OHD and are in ng/ml. This is important because some laboratories measure vitamin D in ng/ml, while others measure in umol/l. The two are not interchangeable, so the units must be considered when interpreting results for plasma vitamin D. A notable outlier is the unusually high vitamin D level seen in headstarted Anegada iguanas. Although such levels may be normal, if they are artificially high, this may result from inadequate shade or excessive competition for shady spots within the headstart facility.

Human Causes of Morbidity

Man-made factors have a direct negative health effect on many free-ranging populations of rock iguanas. These include predation by domestic dogs and cats, predation by non-native predator species, habitat degradation, non-native vegetation out-competing native forage, competition for food resources with goats and cattle, road casualties, and others. The primary factors impacting each population need to be evaluated and addressed on a species-by-species basis.

PATHOLOGY

The best way to increase our knowledge of pathologic processes affecting rock iguanas is to collect data through complete post-mortem evaluations. To date, there are limited data on the pathology of rock iguanas, especially from wild and headstarted populations. Information on the disease processes that affect these populations is lost when pathologic data are not collected. Biologists and veterinarians working with rock iguana conservation programs are encouraged to train range country professionals in the importance of, and techniques for, thorough post-mortem examination. Such efforts can provide a wealth of knowledge relevant to improving the health and well-being of rock iguana populations.

CHAPTER

7

Conservation

CHAPTER

7

Conservation

OUTLINE

Introduction 175

Threats 178

Conservation Actions 185

Outreach and Education 196

Long-Term Species Recovery Planning 199

INTRODUCTION

As a group, West Indian rock iguanas are among the most endangered lizards in the world (Alberts, 2000). Five taxa are ranked as Critically Endangered on the IUCN Red List of Threatened Species (Turks and Caicos, Jamaican, Grand Cayman Blue, Anegada, and Ricord's iguanas), and one is ranked as Endangered (San Salvador iguana). Three are ranked as Vulnerable (Rhinoceros, Bahamian, and Cuban iguanas), but the IUCN Iguana Specialist Group has made a strong recommendation that the Rhinoceros iguana be upgraded to Endangered due to the continuing decline of wild populations. All nine species are afforded the highest level of protection, Appendix I, under CITES, the Convention on International Trade in Endangered Species of Wild Fauna and Flora.

One species, the Jamaican iguana, was believed to have been driven to extinction in the 1940s, until a small remnant population was discovered in the rugged Hellshire Hills in 1990 (Vogel, 1994). Although the population has since grown to over 200 individuals, it is still intensively managed and remains in peril from a multiplicity of threats (Vogel et al., 1996; Wilson et al., 2004a). Two other species, the Anegada and Grand Cayman Blue iguanas, also number no more than a few hundred adults in the wild and are dependent on active conservation management programs for their survival (Table 7.1).

The islands of the West Indies support human population densities that are among the highest anywhere outside Southeast Asia, and the region's population is expected to increase

Cyclura, First Edition DOI: 10.1016/B978-1-4377-3516-1.10007-X

to up to 60 million people by 2025 (Ottenwalder, 2000c). That such a diverse array of ethnicities, languages, and cultures is represented, combined with a rapid rate of urbanization, has made regional conservation especially challenging. Most development, particularly for tourism, is concentrated in the coastal areas that many iguana species depend upon for

TABLE 7.1 Current Wild Adult Population Estimates and Relative Degree of Severity of Current Threats (Low, Moderate, High, Severe) to Rock Iguanas*

Taxon	Current wild population estimate	Habitat loss and degradation	Predation by introduced mammals	Hunting and poaching	Human interactions
Turks and Caicos iguana *Cyclura carinata carinata*	50,000	High	Severe	Low	High
Jamaican iguana *Cyclura collei*	150	Severe	Severe	Low	Low
Rhinoceros iguana *Cyclura cornuta cornuta*	17,000	High	High	Moderate	Moderate
Mona Island iguana *Cyclura cornuta stejnegeri*	1,500	Moderate	Severe	Low	Moderate
Andros Island iguana *Cyclura cychlura cychlura*	3,500	Moderate	Severe	High	Low
Exuma Island iguana *Cyclura cychlura figginsi*	1,500	Low	Moderate	Moderate	Moderate
Allen Cays iguana *Cyclura cychlura inornata*	1,000	Low	Low	Moderate	High
Grand Cayman Blue iguana *Cyclura lewisi*	300	Severe	Severe	Low	High
Sister Isles iguana *Cyclura nubila caymanensis*	1,500	Moderate	High	Low	Moderate
Cuban iguana *Cyclura nubila nubila*	40,000	High	High	Moderate	Moderate
Anegada Island iguana *Cyclura pinguis*	400	High	Severe	Low	Low
Ricord's iguana *Cyclura ricordii*	1,300	High	Severe	Moderate	Low
White Cay iguana *Cyclura rileyi cristata*	200	Low	Low	Moderate	Low
Acklin's iguana *Cyclura rileyi nuchalis*	13,000	Low	Low	Low	Low
San Salvador iguana *Cyclura rileyi rileyi*	500	Low	High	Moderate	Moderate

* *Note that historical threats may differ, particularly in the Bahamas where predation by introduced mammals and hunting by native peoples were likely more important in the past than they are today.*

high quality nesting habitat. When iguanas are also exposed to non-native species such as feral cats and dogs, the results can be devastating (Iverson, 1978).

In general, the tropical dry forests inhabited by iguanas are among the most endangered ecosystems in the world (Murphy and Lugo, 1986) (Figure 7.1). The loss or reduction of rock iguana populations has potentially serious consequences for the health of these thorn forest habitats. Not only are these iguanas important seed dispersers for the dozens of native plant species they consume (Iverson, 1985), but research has also shown that seeds that pass through the digestive tract of an iguana will germinate more rapidly than those that have not (Hartley et al., 2000). Iguana scat also appears to function as a natural fertilizer, resulting in more rapid seedling growth than would otherwise occur (Alberts, 2004). Finally, shoot and foliage production may be enhanced by the ongoing cropping of native plants by iguanas (Knapp and Hudson, 2004).

Despite the fact that most rock iguana populations have suffered serious declines, there is reason to be optimistic about their future. The IUCN's 1997 decision to sanction the formation of a West Indian Iguana Specialist Group (since expanded to become a global Iguana Specialist Group) was instrumental in bringing together experts internationally and throughout the region to develop a coordinated plan for the long-term conservation of West Indian iguanas (Alberts, 2000). A few years later, the International Iguana Foundation (IIF) was founded. The IIF has been critical in raising the profile of iguana conservation and raising funds to support iguana survival worldwide. Together, these groups have supported conservation of iguana habitat, field research, translocation and reintroduction programs,

FIGURE 7.1 Cuban iguana in tropical dry forest habitat.

educational outreach, and the establishment of managed zoo-based populations as a hedge against extinction in the wild.

THREATS

Habitat Loss and Degradation

The primary threats to the long-term survival of rock iguana populations continue to be habitat loss and degradation caused by people and domestic livestock. These effects are ubiquitous throughout the range of rock iguanas, although some good quality iguana habitat has been spared, particularly in parts of Cuba and the Bahamas that are remote, sparsely populated, and lack natural fresh water sources (Ehrig, 2000). Habitat loss takes a variety of forms, including limestone mining, which has destroyed large tracts of habitat on Cuba, Puerto Rico, and Jamaica. Even when the areas directly impacted by mining are limited, the roads that are cut to facilitate mining operations allow further incursion into forest habitats. Other direct sources of habitat loss include agriculture, especially in the Dominican Republic and on Grand Cayman, as well as land clearing for tourist resorts and housing developments (Figure 7.2). In the Dominican Republic, it has been estimated that 35% of Rhinoceros iguana habitat has been lost, and that 75% of that which remains is highly disturbed (Ottenwalder, 2000a).

Although less direct than the impacts of land clearing, habitat fragmentation and disturbance are equally devastating to rock iguana populations. Hardwood timber extraction, especially prevalent on the larger islands such as Cuba and Hispaniola, has disturbed the forest canopy and impacted local watersheds. Wood cutting for the production of charcoal, a cheap cooking fuel, has significantly degraded iguana habitat in Jamaica and the Dominican Republic. Harvest methods traditionally involving machetes have been replaced with chainsaws, greatly increasing the pace of destruction (Wilson et al., 2004b). The Hellshire Hills, last stronghold of the Jamaican iguana, has been severely impacted by charcoal burning activities in the northeast, where today only barren ground remains (Vogel, 2000).

Finally, the feral livestock that invariably accompany human settlement have had serious negative consequences for rock iguana populations (Figure 7.3). Goats, burros, donkeys, sheep, and cattle compete with iguanas for food, alter local vegetation composition, and prevent the regeneration of native plants (Mitchell, 1999). Goats, introduced to the Caribbean in the sixteenth century as a food source for shipwrecked sailors, are particularly destructive, removing high quality iguana food plants through over-browsing. As a result of the disturbance caused by grazing, many iguana habitats have been invaded by invasive plant species such as Australian pine (*Casuarina equisetifolia*), which dominates degraded dry forests throughout the Caribbean. Feral livestock not only cause a shift in plant species composition toward toxic, non-palatable species, but also trample iguana nests and burrows, causing either direct collapse or compaction of soils such that they are unsuitable for future nesting.

Predation by Introduced Mammals

In addition to habitat loss and degradation, predation by introduced mammals is decimating many rock iguana populations. Although the introduced Indian mongoose

FIGURE 7.2 Core iguana habitat in the British Virgin Islands (*a*) which was cleared for development that never occurred (*b*).

FIGURE 7.3 Cattle and other livestock degrade iguana habitat by browsing on native vegetation and trampling suitable nesting habitat.

is currently only known to be a significant problem for one species, the Jamaican iguana, its impact has been devastating (Figure 7.4). Mongooses were first introduced to Jamaica in 1872 in a futile attempt to control black rats (Tolson, 2000). Iguana populations plummeted to the point where the species was believed extinct when it disappeared from the Goat Islands in the mid-1940s. Although a tiny remnant population was rediscovered in Jamaica's Hellshire Hills in 1990, it has taken years of intensive mongoose trapping and removal for the population to show the beginnings of recovery (Wilson et al., 2004a, 2004b; Lewis et al., 2011).

Feral cats and domestic dogs are also a major threat to iguana populations throughout the region (Figure 7.5). While dogs are capable of predating adults of even the largest species, feral cats are more likely to threaten hatchlings and juveniles. However, for iguana species in the Bahamas and Turks and Caicos Islands, feral cats have been shown to kill and consume all age classes. On Pine Cay, a thriving population of more than 5,000 Turks and Caicos iguanas was driven nearly to extinction in less than five years after a few cats and dogs were brought to the island during hotel construction (Iverson, 1978). The population of iguanas on Anegada island is currently far below carrying capacity, likely largely due to the very large and uncontrolled population of feral cats (Bradley and Gerber, 2005). The natural range of the Turks and Caicos iguana has been reduced by 95%, and its current area of occupancy is inversely correlated with the presence of introduced mammals (Gerber and Iverson, 2000). Because rock iguanas evolved in the absence of any mammalian predators, they have no natural defenses against these enemies.

FIGURE 7.4 The Indian mongoose is known to be a significant predator of the Jamaican iguana.

While less destructive to adult iguanas, rats and feral pigs can still have devastating effects on iguana nests, especially immediately following egg deposition when olfactory cues can be strong (Wiewandt and Garcia, 2000). The impact of feral pigs has been especially problematic on Mona Island, Andros Island, and parts of Cuba. There is some

FIGURE 7.5 Ferals cats prey heavily on juvenile iguanas and can kill adult iguanas of smaller species.

evidence that rats may also depress adult iguana densities for smaller species such as the San Salvador iguana (Hayes et al., 2004). In 1996, a single introduced raccoon decimated the breeding population of iguanas on White Cay in the Bahamas (Hayes, 2000a). It was hypothesized that females were disproportionately impacted because of their greater visibility and the energetic demands placed on them during the nesting season (Hayes et al., 2004).

Hunting and Poaching

Hunting is a less severe threat to rock iguanas than habitat loss or predation by introduced mammals, but it is still a serious danger for some taxa. The hunting that does occur takes two very different forms: local subsistence hunting for food and deliberate poaching for the illegal pet trade. In rural areas of both Haiti and the Dominican Republic, Rhinoceros and Ricord's iguanas are hunted locally for food (Ottenwalder, 2000a, 2000b). In addition, iguana remains have been found in abandoned fishing camps on Andros Island in the Bahamas (Knapp, 2005b) (Figure 7.6). For two taxa of Bahamian iguanas, the Allen's Cay iguana and the White Cay iguana, incidences of in-country poaching and illegal smuggling to the United States for the pet trade have been reported, respectively (Iverson, 2000; Hayes, 2000a).

In 2008, a senseless tragedy at the Queen Elizabeth II Botanic Park iguana breeding facility on Grand Cayman illustrated the most serious single incidence of illegal loss of life on record (Binns, 2008). Vandals broke into the facility under cover of night and deliberately killed

FIGURE 7.6 Remains of an iguana discovered at a hunting camp on the West Side of Andros Island, Bahamas. *Photo by Chuck Knapp.*

FIGURE 7.7 Digger, a large and friendly Grand Cayman Blue iguana, was one of the unfortunate victims of a brutal and deliberate slaughter of captive blue iguanas at the Blue Iguana Recovery Program facility on Grand Cayman.

seven adult Grand Cayman Blue iguanas, representing 17.5% of the captive population, including several important genetic founders (Figure 7.7). Two of the animals that were attacked managed to survive, and the incident resulted in an outpouring of local, national, and international support for the program. Nevertheless, humans clearly remain some of the most potent predators of rock iguanas today.

Human Interactions

Human recreational activities tend to be concentrated on beaches, which is often where iguanas nest communally. Walking through nest clearings can cause nest chambers to collapse and damage developing eggs. An emerging threat comes from well-meaning visitors who feed iguanas on tourist beaches, leading to high levels of intraspecific aggression and disruption of natural social systems. In a study of Cuban iguanas on the US Naval Base at Guantanamo Bay, Lacy and Martins (2003) found that iguanas inhabiting areas with high human interference showed increased levels of male–male aggression and fewer males interacting with females, both of which could significantly impact the natural mating system of this species (Figure 7.8).

FIGURE 7.8 At Guantanamo Bay, Cuba, iguanas with high human interference show increased levels of intraspecific aggression, with fewer males interacting with females.

In the Bahamas, powerboat tours advertising the opportunity to feed wild iguanas often visit the same cays multiple times in a single day. In a study designed to assess the impact of this practice, iguanas were compared on two islands without visiting tourists and three islands where tourists visit regularly and feed iguanas, often with inappropriate food items such as bread, cake, and potato chips. Blood samples collected immediately after capture showed differences in glucose, sodium, hemoglobin, packed cell volume, total solids, and overall body condition (K. Hines and C. Knapp, unpublished data). These data still remain to be analyzed, but it is clear that artificial feeding can result in detectable changes in the physiology of wild iguanas that potentially have a significant impact on their survival.

If they are carefully designed and monitored, human–iguana interactions may be managed in a manner that benefits the conservation of wild iguana populations. On Little Water Cay in the Turks and Caicos Islands, visitation is regulated and controlled by the Turks and Caicos National Trust. The Trust oversaw construction of a boardwalk that ensures that visitors do not inadvertently walk over iguana burrows or trample vegetation. Signage highlights aspects of the biology of Turks and Caicos iguanas, provides information on their conservation status, and gives details of appropriate iguana etiquette. Iguana-appropriate food items are provided for tourists to offer, and each visitor is charged a modest fee that is used to help fund iguana conservation and education programs.

Because rock iguanas are large, photogenic, and charismatic, they can play a role as important flagship species that promote conservation of their threatened dry forest habitats. Ecotourism, if carefully and appropriately managed, can be a valuable means to engage the public in conservation efforts. In particular, citizen science programs such as that managed by the John G. Shedd Aquarium in the Bahamas have strong potential to successfully engage volunteers and garner long-term support (Knapp, 2004).

FIGURE 7.9 This Cuban iguana met an unfortunate end when it was killed by a car.

Every year, many iguanas are killed on roads by vehicular traffic (Figure 7.9), especially in the Cayman Islands, where tourism is rapidly accelerating and few untouched natural areas remain (Burton, 2004b, 2010). Coastal roads can be especially hazardous for species that must successfully navigate across these roads in order to reach the sandy beach habitat required for nesting. Iguana-crossing signage and bumper stickers ("Give Iguanas a Brake") are being explored as a potential deterrent to this problem at key sites, but the effectiveness of this approach remains to be quantified (C. Knapp, personal communication).

CONSERVATION ACTIONS

Habitat Protection

Habitat protection is critical to the survival of all rock iguana populations, although the need is particularly urgent for the Anegada iguana (Figure 7.10). In 2008, a System Plan for Protected Areas for the British Virgin Islands received government approval. The plan includes protection of critical iguana habitat on Anegada Island (J. Smith-Abbott, personal communication). Full implementation awaits resolution of land title issues, but this is an important first step in setting aside protected habitat for this critically endangered species.

In the Dominican Republic, about 60% of the habitat occupied by Ricord's iguanas has been protected within Lago Enriquillo National Park (Ottenwalder, 2000b). Outside protected areas, however, land conversion for agriculture threatens key habitats. In inland areas where nesting sites are scarce, fondos, or soil depressions within rocks, are critical for egg laying. In 2004, the government created a Municipal Protected Area in the

FIGURE 7.10 Critical coastal (*a*) and inland (*b*) habitat on Anegada that awaits government protection.

Pedernales region that covers all four of the major known fondos (Rupp et al., 2008). Despite this, an incident occurred in 2006 that involved destruction of more than 40 iguana nests in one of the fondos by bulldozers, which resulted in the need to fence the area against further intrusion. In an effort to control the pace of development and keep their natural resources intact, the town council of Perdernales voted to become a "Town of Sustainable Tourism" (Rupp et al., 2005).

Although much primary rock iguana habitat, particularly in coastal areas, remains unprotected, progress is being made. In direct response to the conservation needs of the Andros Island iguana, the Bahamas National Trust successfully expanded the Andros national park system to include new areas of South Andros essential to the iguana's survival (IUCN Iguana Specialist Group, 2008). In Grand Cayman, the creation of the Salina Reserve was instrumental in the initial effort to re-establish a wild population of iguanas. The reserve was chosen as a reintroduction site in 2004, and since then more than 260 Grand Cayman Blue iguanas have been released and the population continues to grow (Burton, 2010). Prior to the creation of the reserve, iguanas existed only in isolated pockets of habitat and on the grounds of the Queen Elizabeth II Botanic Park.

Another critical area of iguana habitat is the Goat Islands of Jamaica, located just south of the Hellshire Hills. Even when Jamaican iguanas had been extirpated from the vast majority of their native range, the population on Great Goat Island persisted into the late 1940s (Vogel, 2000). At present, the persistence of the only known extant Jamaican iguana population, which inhabits the Hellshire Hills, is dependent on continued predator control programs. The Goat Islands represent an unparalleled opportunity to re-establish a population of this species in a portion of its native range in which it would be possible to eradicate introduced mammals, precluding the need for ongoing control. The islands support a sizeable population of goats, which would require removal prior to iguana reintroduction, but sufficient native vegetation remains for the habitat to be capable of full recovery (Wilson et al., 2004b). In addition, good quality iguana nesting habitat is present. However, despite the fact that the Goat Islands are included in the Portland Bight Protected Area, management authority for the islands continues to be in flux and their future as an iguana sanctuary remains uncertain.

Invasive Species Control

The Jamaican iguana is currently the only species for which a large-scale, coordinated invasive species control effort is in place. For more than a decade, a field research team from the University of the West Indies has been actively trapping mongooses and feral cats in the core iguana habitat of the Hellshire Hills (Wilson and van Veen, 2006). The size of the resident iguana breeding population has more than doubled since mongoose control efforts were initiated (Lewis et al., 2011). In addition to mongoose trapping, the team also actively controls feral pigs, cats, and dogs that intrude into the core iguana habitat. This effort has been instrumental in ensuring the survival of the juvenile Jamaican iguanas that have been repatriated as part of an active headstarting strategy.

There has also been some limited invasive species control in the Turks and Caicos Islands. These efforts include feral cat trapping on Long Cay (N. Mitchell, personal communication) and the removal of rats from several smaller cays prior to large-scale translocations aimed at re-establishing iguana populations (Gerber, 2007). In addition, a concerted trapping effort is currently under way on Little Water Cay to prevent the establishment of feral cats on this iguana sanctuary. Previously, Little Water had been isolated from the adjacent Pine-Water Cay complex, which unlike Little Water, is inhabited by people. However, the natural emergence of a sand bridge allowed cats to cross onto Little Water Cay, where they are threatening the local iguana population (Figure 7.11). Currently, dredging of a sea channel and construction of a fence are being explored as possible alternatives to keep Little Water Cay predator-free (G. Gerber, personal communication).

On Mona Island in Puerto Rico, hunting is permitted, which helps to control local populations of feral pigs and goats. However, there is pressure from the hunting community to allow these invasive species to continue to persist, which poses challenges for the successful conservation of Mona's unique iguanas. Fortunately, an active fencing program managed by the Department of Natural Resources and the Environment is in place that excludes invasive species from the most sensitive areas, including nesting depressions (Wiewandt and Garcia, 2000).

Even in protected settings, iguanas can still be at risk. Dogs have killed captive adult blue iguanas in their enclosures at the Queen Elizabeth II Botanic Park on Grand Cayman, and the juveniles that roam free in the Park are at risk from feral cats. The Blue Iguana Recovery Program

FIGURE 7.11 A female Turks and Caicos iguana basks next to a fresh set of cat tracks.

maintains an active trapping effort to help ensure the survival of released iguanas in the Park. In addition, the provision of artificial retreats has proven instrumental in giving released iguanas a place to hide and decreasing losses due to predation by feral mammals (Burton, 2010).

A new partnership between the IUCN Iguana Specialist Group and the NGO Island Conservation holds great promise. Island Conservation specializes in ecosystem restoration through invasive species removal and has successfully eradicated cats and rats from several islands off the coast of Southern California and Mexico in order to facilitate the re-establishment of seabird colonies and endemic plants. Island Conservation is expanding their work to include the Caribbean islands and several projects, including the potential to eradicate feral cats from Anegada Island and feral goats from Jamaica's Goat Islands, are being discussed and analyzed.

Headstarting

Headstarting is a conservation strategy that consists of rearing juveniles in a safe captive environment until they are large enough to survive on their own in the wild (Figure 7.12). Research has shown that rock iguanas are good candidates for headstarting because many behaviors directly related to survival appear to be innate (Alberts, 2007). Experimental studies showed that juvenile Cuban iguanas headstarted for 18 months at the San Diego Zoo showed no decrease in the intensity of their response to potential predators, no reluctance to feed on natural food sources despite having been on an artificial diet, and no deficits in normal social interactions following release (Alberts et al., 2004a).

Because young rock iguanas are so vulnerable to predation by introduced species, headstarting has been instrumental to the survival of several species. In particular, the Jamaican

FIGURE 7.12 A group of headstarted Cuban iguanas at the San Diego Zoo.

iguana appears to have been rescued from the brink of extinction by a combination of headstarting and aggressive predator control (Wilson et al., 2004a). When a tiny remnant population of Jamaican iguanas was rediscovered clinging to survival in the Hellshire Hills in 1990, immediate action was taken to protect the only known nesting sites, rescue hatchlings from wild nests, and transport them to Kingston's Hope Zoo as part of a coordinated headstarting effort (Vogel et al., 1996) (Figure 7.13). Since then, over 140 headstarted Jamaican iguanas have been released into the Hellshire Hills. Radiotelemetry studies indicate that headstarted females now make up 40% of females nesting in the wild, demonstrating their successful integration into the wild breeding population and highlighting the key contribution that headstarting is making to species survival (B. Wilson and R. van Veen, unpublished data).

Headstarting has also been an integral part of the recovery effort for the Grand Cayman Blue iguana. The first captive facility was built at the Queen Elizabeth II Botanic Park in 1996, and rapidly expanded thereafter. By 2004, the program had reached its goal of supplying 100 two-year-old iguanas for release annually (Burton, 2005). Releases onto Park grounds resulted in the establishment of relatively stable home ranges and a small breeding population. In December 2004, the first releases into natural habitat occurred, when 23 headstarted iguanas were reintroduced to the Salina Reserve, a 625-acre protected area managed by the National Trust for the Cayman Islands (Burton, 2005). Survival at both release sites has been high, likely due in part to the provision of artificial wooden burrows that served to anchor young iguanas to their release sites and presumably reduced mortality associated with excessive movement following release. Extensive radiotracking

FIGURE 7.13 The headstart facility at the Hope Zoo in Kingston, Jamaica.

FIGURE 7.14 Headstart facilities such as this one on Anegada have been very successful in bolstering wild iguana populations.

showed that although most headstarted iguanas lost weight following release, 91% of the first season's released animals were still alive seven months later.

In 1997, headstarting was instituted for a third species for which the wild population was precariously low, the Anegada iguana (Figure 7.14). Together, the British Virgin Islands National Parks Trust and the IUCN Iguana Specialist Group constructed an on-island breeding and rearing facility that also serves as an informal education center for the public. Releases of four- to six-year-old iguanas on Anegada have been carried out as part of a controlled experimental design aimed at determining the smallest body size at which an iguana can resist cat predation (Bradley and Gerber, 2005). Initial results indicated that animals as small as 0.55–0.75kg can still survive in the presence of feral cats. Survival rates approached 90%, suggesting that even younger, smaller age classes can likely be successfully released in the future.

Building on these earlier successes, a headstarting facility was established on Mona Island in Puerto Rico in 1999 for the endangered Mona Island iguana (Garcia et al., 2007; Perez-Buitrago et al., 2008). To date, releases have been conducted in an experimental context in an effort to evaluate their effectiveness and guide future improvements to headstart-release methods. Headstarting did appear to positively influence survival rates in the wild, with a 40.3% minimum survival rate for headstarted hatchlings compared to 22% for wild hatchlings. At least two of the released females were documented to have bred in the wild following release (Garcia et al., 2007). Dispersal was variable and erratic, with many individuals eventually migrating back from the release point to the rearing facility. Future studies are needed to determine the implications of this atypical dispersal pattern for long-term population management. Another key finding was that while growth of headstarted iguanas did not differ from that of their wild counterparts, there was a significant decrease in growth rates in the third year of captivity that was likely a result of overcrowding. Assuming that post-release survival rates are not impacted, this suggests that it may be prudent to limit the headstarting period to two years.

Integral to all of these programs has been a rigorous health screening regimen designed to ensure that all headstarted animals are free of diseases or parasites that could be transmitted to wild populations. The Fort Worth Zoo, the Wildlife Conservation Society, and others have provided critical veterinary support to headstarting programs in Jamaica, Grand Cayman, and the British Virgin Islands. To date, no significant disease findings have emerged, suggesting that current husbandry protocols are sufficient to maintain healthy colonies and support safe release practices.

Translocation Programs

Another strategy for increasing the number and size of wild populations involves moving groups of iguanas into new unoccupied habitat via translocation. Translocation may be particularly well-suited to situations in which rock iguanas are restricted to few or small islands that may be especially vulnerable to chance extinction events (Knapp and Hudson, 2004). The largest translocation effort to date has been in the Turks and Caicos Islands (Figure 7.15). In 2002 and 2003, researchers from the San Diego Zoo relocated iguanas from two islands where they were threatened by development and feral cat predation to four protected cays in the region from which iguanas had been extirpated (Gerber, 2007). Depending on its size and estimated carrying capacity, each recipient cay received between 18 and 182 iguanas. In the five months immediately following release, iguanas showed some signs of dehydration and stress, but thereafter were as healthy as their wild counterparts. Annual post-release monitoring has shown that iguanas hatched on the recipient cays have grown two to three times faster than similarly aged iguanas on source islands. Remarkably, the translocated iguanas have also reached sexual maturity in as early as two years, compared to a typical maturation rate of six or seven years.

Translocation has also been used extensively as a conservation strategy in the Bahamas. The current natural population of the Allen Cays iguana occurs on only two small Cays, Leaf and Southwest Allan's. In any effort to create an additional population as a hedge against extinction in the wild, four male and four female iguanas were translocated to Alligator Cay between 1988 and 1990 (Knapp, 2001). Seven of these individuals were known to have survived and served as the founders for a growing self-sustaining population. Genetic studies at 10 years post translocation showed that each of the founding males contributed equally to the population and that genetic diversity was largely retained (Knapp and Malone, 2003). Similar to the Turks and Caicos Island translocations, growth rates of translocated iguanas were significantly higher than those on source islands. It is likely that low population density, abundant food supply, and unrestricted nest sites all contributed to high reproductive output and rapid growth of translocated iguanas. Given that high rates of human

FIGURE 7.15 Dr. Allison Alberts and Dr. Glenn Gerber (San Diego Zoo Global) prepare to release a pair of Turks and Caicos iguanas as part of a large-scale translocation project in the Turks and Caicos Islands.

visitation are impacting both source populations, the relatively isolated and protected nature of Alligator Cay will presumably enhance the long-term security of the species.

Several translocated populations of Anegada iguanas have also been established, beginning with the removal of eight individuals from Anegada to found a new population on Guana Island in the mid-1980s (Goodyear and Lazell, 1994). The Guana population continues to thrive (Figure 7.16), with individual iguanas showing higher body weights than their counterparts on Anegada, likely due to a greater abundance of food plants (Perry et al., 2007). Guana has since served as the source population for additional translocations to several other small, privately owned islands in the region.

Because they are adapted to harsh environmental conditions and many of the behaviors critical for survival are innate, rock iguanas appear to be excellent candidates for translocation. Of critical importance is identifying islands suitable for establishing new populations. However, if recipient islands are free of introduced predators and human disturbance, sufficient foraging and nesting habitat are available, and suitable retreats exist or can be created, thriving populations can grow from a modest number of founders in just a few years.

The Role of Zoos and Other Conservation Organizations

In addition to public education, which forms an essential cornerstone of the mission of any zoo, the two primary avenues by which zoos have contributed to rock iguanas are through

FIGURE 7.16 Anegada Island iguana on Guana Island.

fund raising and captive breeding. Without adequate funding, conservation is not possible, and it takes a substantial investment to support research and recovery efforts. The role of zoos in the conservation of rock iguanas has been reviewed by Hudson and Alberts (2004), but it is worth reiterating the critical financial and logistical contributions that zoos have made to the construction and maintenance of headstarting facilities on Jamaica, Grand Cayman, and Anegada Island.

The International Iguana Foundation (IIF) has done significant fund raising for both in-situ and ex-situ conservation programs. The board of the IIF is primarily made up of zoo conservationists and researchers whose respective institutions make an annual donation to directly support the conservation work of the foundation. Additional funds are raised through public and private donations, as well as grants. Another group that has raised substantial amounts of funding for iguana conservation is the International Reptile Conservation Foundation (IRCF). IRCF has provided critical support for many programs, especially the Blue Iguana Recovery Program in Grand Cayman, including a very active volunteer program.

Over time, captive reproduction of rock iguanas has become more common in zoos. The primary goal of these programs is to preserve the genetic diversity of the most critical wild populations and provide offspring for reintroduction in the case of natural or human-caused catastrophes. Zoo populations of iguanas are carefully managed under the auspices of studbooks that are maintained by a professional from an Association of Zoos and Aquariums (AZA) institution. The AZA Rock Iguana Species Survival Plan (SSP) was

FIGURE 7.17 The Grand Cayman Blue iguana is one of three critically endangered rock iguanas for which a Species Survival Plan has been implemented.

approved in April 1996. Although all 15 taxa comprising the genus *Cyclura* are threatened, the SSP is focused on three of the most critically endangered: the Jamaican, Grand Cayman Blue, and Anegada iguanas (Figure 7.17). The goals of the SSP are to manage captive populations as a hedge against extinction in nature, and to utilize zoo-based programs to generate support for ongoing field research and recovery programs. Nineteen AZA zoos and aquariums have signed a memorandum of understanding for this SSP, and over 20 have contributed funds to support conservation initiatives for most of the taxa, including all of those ranked as Critically Endangered. At least nine SSP institutions are directly participating with field research and conservation initiatives in eight Caribbean countries.

Scientific Research

Understanding the biology and habitat requirements of rock iguanas through field research is essential to developing effective conservation and recovery programs. To varying degrees, field research has been carried out on all taxa of rock iguanas, with much of it

FIGURE 7.18 Andros iguana with attached radio transmitter sutured through the dorsal crest. *Photo by Chuck Knapp.*

directed toward conservation of the most critically endangered species (Alberts et al., 2004b). New insights have been gained into social behavior and communication, reproduction, foraging ecology, health and nutrition, and spatial movement patterns. Radiotelemetry (Figure 7.18) has been especially important in providing new understanding into the biology of rock iguanas in their native habitats (Goodman et al., 2009).

Most of the current scientific research on rock iguanas builds on the early work of two doctoral dissertations conducted by IUCN Iguana Specialist Group founding members Dr. John Iverson (Iverson, 1979) and Dr. Tom Wiewandt (Wiewandt, 1977). Both of these pioneering studies elucidated many new aspects of rock iguana natural history, and set the stage for much of the work that followed in ensuing decades. As marking and monitoring techniques have improved, valuable new information about nesting ecology, seasonal movement patterns, and human–iguana interactions has emerged. Much of the recent research on rock iguanas has been directed toward the development of science-based management strategies, including population monitoring, conservation genetics, and reintroduction and translocation biology (Alberts et al., 2004b).

OUTREACH AND EDUCATION

Outreach and education are essential to the success of conservation efforts for rock iguanas. When carefully implemented in cooperation with local educators and managers, education can be critically important to instilling local pride and admiration for iguanas.

Virtually all rock iguana conservation programs currently contain significant education components (Figure 7.19). The headstarting facilities on Grand Cayman and Anegada are open to the public, providing an opportunity for visitors to learn about the threats facing these unique species and the efforts currently under way to recover and conserve wild populations. As part of a special program, every Friday hotel guests at the Ritz–Carlton are given the opportunity to volunteer their time to help the Blue Iguana Recovery Program. At the Hope Zoo in Kingston, Jamaican iguanas are displayed proudly with the appropriate slogan "Nuff Respect Due."

Reaching out to children in an effort to instill a conservation ethic remains a hallmark of iguana conservation initiatives, perhaps nowhere more so than on Andros Island in the Bahamas (Knapp, 2005b). Programs include offering high school students the opportunity to assist in the field, giving presentations to schools and community groups, murals and drawing contests, and the establishment of a local soccer club, the Central Andros Iguanas.

Programs aimed at students and teachers have been particularly successful in the Dominican Republic (Rupp et al., 2005). Grants from the US Fish and Wildlife Service and the Association of Zoos and Aquariums funded the development of curriculum materials

FIGURE 7.19 Rock iguana researchers from the San Diego Zoo educate visitors about the iguanas in the Turks and Caicos Islands.

focused on iguana natural history and conservation, as well as teacher training workshops in the capital city, Santo Domingo, and in outlying areas where iguanas naturally occur. North American zoos were also instrumental in developing teacher training workshops in the Turks and Caicos Islands.

The International Reptile Conservation Foundation, the IUCN Iguana Specialist Group, and the International Iguana Foundation have worked together to produce a series of educational posters, newsletters, and brochures ranging from general information about the conservation status of rock iguanas, to specific issues such as the negative effects of introduced green iguanas. In the Turks and Caicos Islands, the tourist boardwalk on Little Water Cay includes a number of educational displays focused on iguana biology and conservation.

Despite these efforts, more still needs to be done. In particular, training and capacity building for local wildlife managers is needed to share the latest techniques in population monitoring, habitat management, captive husbandry, and veterinary care. This is especially important for those species for which emergency headstarting efforts are already under way. In a model that could be replicated at other locations, the Jamaican iguana recovery program

FIGURE 7.20 Cuban rock iguana at sunset.

provided hands-on training opportunities to several students from the University of the West Indies, resulting in a growing capacity for in-country field research.

LONG-TERM SPECIES RECOVERY PLANNING

In 1994, rock iguana experts from around the world convened in Kingston, Jamaica, to participate in a Population and Habitat Viability Workshop. The impetus for the workshop was the astounding rediscovery that a small remnant population of the Jamaican iguana, previously believed extinct, had managed to persist. A recovery plan was developed, leading to initial field studies and the inception of the headstarting program at the Hope Zoo. In addition, this milestone workshop formed the basis for the formation of the IUCN Iguana Specialist Group and ultimately resulted in the publication of the IUCN's first Action Plan for West Indian Iguanas (Alberts, 2000). Since then, the Specialist Group has worked with a variety of partners to develop conservation management and species recovery plans for the Grand Cayman Blue (2001), Ricord's (2002), Turks and Caicos (2003), Anegada (2004), Andros (2005), and Jamaican (2006) iguanas. These plans have served to bring together local stakeholders to develop a common vision and catalyze national and international support for the long-term survival of rock iguanas. In 2010, the group met in Cuba to exchange key information on the biology and conservation of Cuban iguanas (Figure 7.20).

Bibliography

Adams, C. (2004). Obituary. *Iguana Specialist Group newsletter, 7*(1), 2.

Alberts, A. C. (1993). Chemical and behavioral studies of femoral gland secretions in iguanid lizards. *Brain, Behavior, and Evolution, 41*, 255–260.

Alberts, A. C. (1995). Use of statistical models based on radiographic measurements to predict oviposition date and clutch size in rock iguanas (*Cyclura nubila*). *Zoo Biology, 14*, 543–553.

Alberts, A. C. (2000). *West Indian Iguanas: Status Survey and Conservation Action Plan*. Gland, Switzerland: IUCN – The World Conservation Union.

Alberts, A. C. (2004). Conservation strategies for West Indian rock iguanas (genus *Cyclura*): Current efforts and future directions. *Iguana, 11*, 213–223.

Alberts, A. C. (2007). Behavioral considerations of headstarting as a conservation strategy for endangered Caribbean rock iguanas. *Applied Animal Behaviour Science, 102*, 380–391.

Alberts, A. C., Perry, A. M., Lemm, J. M., & Phillips, J. A. (1997). Effects of incubation temperature and water potential on growth and thermoregulatory behavior of hatchling Cuban rock iguanas (*Cyclura nubila*). *Copeia, 1997*, 766–776.

Alberts, A. C., Oliva, M. L., Worley, M. B., Telford, S. R., Jr., Morris, P. J., & Janssen, D. L. (1998). The need for pre-release health screening in animal translocations: A case study of the Cuban iguana (*Cyclura nubila*). *Animal Conservation, 1*, 165–172.

Alberts, A. C., Grant, T. D., Gerber, G. P., Comer, K. E., Tolson, P. J., Lemm, J. M., & Boyer, D. (2001). *Critical Reptile Species Management on the US Naval Base, Guantanamo Bay, Cuba*: Report to the United States Navy for Project No. 62470-00-M-5219.

Alberts, A. C., Lemm, J. M., Perry, A. M., Morici, L. S., & Phillips, J. A. (2002). Temporary alteration of social structure in a threatened population of Cuban iguanas (*Cyclura nubila*). *Behavioral Ecology and Sociobiology, 51*, 324–335.

Alberts, A. C., Lemm, J. M., Grant, T. D., & Jackintell, L. A. (2004a). Testing the utility of headstarting as a conservation strategy for West Indian iguanas. In A. C. Alberts, R. L. Carter, W. K. Hayes, & E. P. Martins (Eds.), *Iguanas: Biology and Conservation* (pp. 210–219). Los Angeles: University of California Press.

Alberts, A. C., Carter, R. L., Hayes, W. K., & Martins, E. P. (2004b). *Iguanas: Biology and Conservation*. Los Angeles: University of California Press.

Allen, G. M. (1937). *Geocapromys* remains from Exuma. *Journal of Mammalogy, 18*, 369–370.

Allen, M. E., & Oftedal, O. T. (2003). The nutritional management of an herbivorous reptile, the green iguana (*Iguana iguana*). In E. R. Jacobson (Ed.), *Biology, Husbandry, and Medicine of the Green Iguana,* Iguana iguana (pp. 47–74). Melbourne, FL: Krieger.

Allen, M. E., Oftedal, O. T., Baer, D. J., & Werner, D. I. (1989). Nutritional studies with the green iguana. In T. P. Meehan, S. D. Thompson, & M. E. Allen (Eds.), *Proceedings of the Eighth Dr. Scholl Conference on the Nutrition of Captive Wild Animals* (pp. 73–81). Chicago, IL: Lincoln Park Zoo.

Allen, M. E., Chen, T. C., Holick, M. F., & Merkel, E. (1999). Evaluation of vitamin D status of the green iguana (*Iguana iguana*): Oral administration vs. UVB exposure. In M. F. Holick, & E. G. Jung (Eds.), *Biologic Effects of Light* (pp. 99–102). Boston, MA: Kluwer.

Auffenberg, W. (1976). Rock iguanas. Part II. *Bahamas Naturalist, 2*, 9–16.

Auffenberg, W. (1982). Feeding strategies of the Caicos ground iguana, *Cyclura carinata*. In G. M. Burghardt, & A. S. Rand (Eds.), *Iguanas of the World* (pp. 84–116). Park Ridge, NJ: Noyes Publication.

Backeus, K. A., & Ramsay, E. C. (1994). Ovariectomy for treatment of follicular stasis in lizards. *Journal of Zoo and Wildlife Medicine, 25*, 111–116.

Baer, D. J. (2003). Nutrition in the wild. In E. R. Jacobson (Ed.), *Biology, Husbandry, and Medicine of the Green Iguana,* Iguana iguana (pp. 38–46). Melbourne, FL: Krieger.

Baer, D. J., Oftedal, O. T., Rumpler, W. V., & Ullrey, D. E. (1997). Dietary fiber influences nutrient utilization, growth and dry matter intake of green iguanas (*Iguana iguana*). *Journal of Nutrition, 127*, 1501–1507.

Barbour, T. (1917). Notes on the herpetology of the Virgin Islands. *Proceedings of the Biological Society of Washington, 30*, 97–104.

Barbour, T. (1919). A new rock iguana from Porto Rico. *Proceedings of the Biological Society of Washington, 32*, 145–148.

Barbour, T. (1923). Another new Bahaman iguana. *Proceedings of the New England Zoological Club, 8*, 107–109.

Barbour, T. (1937). Third list of Antillean reptiles and amphibians. *Bulletin of the Museum of Comparative Zoology, 82*, 77–166.

Barbour, T., & Noble, G. K. (1916). A revision of the lizards of the genus. *Cyclura. Bulletin of the Museum of Comparative Zoology, 60*, 137–164.

Barus, V., & Coy Otero, A. (1969). Systematic survey of nematodes parasitizing lizards (Sauria) in Cuba. *Helminthologia, 10*, 329–346.

Barus, V., Hubálek, Z., & Coy Otero, A. (1996). Testing the context and extent of host–parasite co-evolution: nematodes parasitizing Cuban lizards (Sauria: Iguanidae). *Folia Zoologica, 45*, 57–64.

Bennett, A. F., & Gorman, G. C. (1979). Population density and energetics of lizards on a tropical island. *Oecologia, 42*, 339–358.

Bernard, J. B. (1995). *Spectral irradiance of fluorescent lamps and their efficacy for promoting vitamin D synthesis in herbivorous reptiles.* PhD Dissertation. East Lansing, MI: Michigan State University.

Bernard, J. B., Oftedal, O. T., Citino, S. B., Ullrey, D. E., & Montali, R. J. (1991). The response of vitamin D-deficient green iguanas (*Iguana iguana*) to artificial ultraviolet light. *Proceedings American Association of Zoo Veterinarians, 1991*, 147–150.

Bernard, J. B., Oftedal, O. T., & Ullrey, D. E. (2006). Idiosyncrasies of vitamin D metabolism in the green iguana (*Iguana iguana*). *Proceedings of the Comparative Nutrition Society Symposium, 2006*, 11.

Binns, J. (2008). The seven blues of May. *Iguana, 15*, 67–77.

Bjourndal, K. A. (1997). Fermentation in reptiles and amphibians. In R. I. Mackie, B. A. White, & R. E. Isaacson (Eds.), *Gastrointestinal Microbiology. Gastrointestinal Ecosystems and Fermentations, Vol. I* (pp. 199–233). New York: Chapman and Hall.

Blair, D. (1991). West Indian iguanas of the genus *Cyclura*: Their current status in the wild, conservation priorities and efforts to breed them in captivity. *Northern California Herpetological Society Special Publication, 6*, 55–66.

Bogoslavsky, B. A. (2007). The use of Ponazuril to treat coccidiosis in eight inland bearded dragons (*Pogona vitticeps*). *Proceedings of the Association of Reptile and Amphibian Veterinarians, 2007*, 8.

Bonnaterre, P. J. (1789). *Tableau Encyclopedique et Methodique des Trois Regnes de la Nature. Erpetologie.* Libraire, Paris: Panckoucke.

Boylan, T. (1984). Breeding the Rhinoceros iguana (*Cyclura c. cornuta*) at the Sydney Zoo. *International Zoo Yearbook, 23*, 144–148.

Bradley, K. A., & Gerber, G. P. (2005). Conservation of the Anegada iguana (*Cyclura pinguis*). *Iguana, 12*, 79–85.

Bryan, J. J., Gerber, G. P., Welch, M. E., & Stephen, C. L. (2007). Re-evaluating the taxonomic status of the Booby Cay iguana, *Cyclura carinata bartschi. Copeia, 2007*, 734–739.

Buckner, S., & Blair, D. (2000a). Andros Island iguana: *Cyclura cychlura cychlura*. In A. C. Alberts (Ed.), *West Indian Iguanas: Status Survey and Conservation Action Plan* (pp. 31–32). Gland, Switzerland: IUCN – the World Conservation Union.

Buckner, S., & Blair, D. (2000b). Bartsch's iguana: *Cyclura carinata bartschi*. In A. C. Alberts (Ed.), *West Indian Iguanas: Status Survey and Conservation Action Plan* (pp. 18–19). Gland, Switzerland: IUCN – the World Conservation Union.

Burton, F. J. (2000). Grand Cayman iguana: *Cyclura nubila lewisi*. In A. C. Alberts (Ed.), *West Indian Iguanas: Status Survey and Conservation Action Plan* (pp. 45–47). Gland, Switzerland: IUCN – the World Conservation Union.

Burton, F. J. (2004a). Revision to species of *Cyclura nubila lewisi*, the Grand Cayman blue iguana. *Caribbean Journal of Science, 40*, 198–203.

Burton, F. J. (2004b). Battling extinction: A view forward for the Grand Cayman Blue Iguana (*Cyclura lewisi*). *Iguana, 11*, 233–237.

Burton, F. J. (2005). Restoring a new wild population of blue iguanas (*Cyclura lewisi*) in the Salina Reserve, Grand Cayman. *Iguana, 12*, 167–174.

Burton, F. J. (2006). Blue iguana recovery program. *Iguana, 13*, 117.

Burton, F. J. (2010). *The Little Blue Book: A Short History of the Grand Cayman Blue Iguana*. San Jose, California: International Reptile Conservation Foundation.

Campbell, T. W. (2006). Hemoparasites. In D. Mader (Ed.), *Reptile Medicine and Surgery* (pp. 801–805). St Louis, MO: Elsevier Saunders.

Carey, W. M. (1966). Observations on the ground iguana *Cyclura macleayi caymanensis* on Cayman Brac, British West Indies. *Herpetologica, 22*, 265–268.

Carey, W. M. (1975). The rock iguana, *Cyclura pinguis*, on Anegada, British Virgin Islands, with notes on *Cyclura ricordi* and *Cyclura cornuta* on Hispaniola. *Bulletin of the Florida State Museum of Biological Sciences, 19*, 189–234.

Carey, W. M. (1976). Iguanas of the Exumas. *Wildlife, 18*, 59–61.

Carpenter, J. W. (2005). *Exotic Animal Formulary* (3rd edn.). St Louis, MO: Elsevier Saunders.

Carrillo-Lopez, A., Yahia, E. M., & Ramirez-Padilla, G. K. (2010). Bioconversion of carotenoids in five fruits and vegetables to vitamin A measured by retinol accumulation in rat livers. *American Journal of Agricultural and Biological Sciences, 5*, 215–221.

Carter, R. L., & Hayes, W. K. (2004). Conservation of an endangered Bahamian rock iguana. II. In A. C. Alberts, R. L. Carter, W. K. Hayes, & E. P. Martins (Eds.), *Iguanas: Biology and Conservation* (pp. 258–273). Los Angeles: University of California Press.

Cerny, V. (1969). The tick fauna of Cuba. *Folia Parasitologica, 16*, 279–284.

Christian, K. (1986a). Physiological consequences of nighttime temperature for a tropical, herbivorous lizard (*Cyclura nubila*). *Canadian Journal of Zoology, 64*, 836–840.

Christian, K. (1986b). Aspects of the life history of Cuban iguanas on Isla Magueyes, Puerto Rico. *Caribbean Journal of Science, 22*, 159–164.

Christian, K., Clavijo, I. E., Cordero-Lopez, N., Elias-Maldonado, E. E., Franco, M. A., Lugo-Ramirez, M. B., & Marengo, M. (1986). Thermoregulation and energetics of a population of Cuban iguanas (*Cyclura nubila*) on Isla Magueyes, Puerto Rico. *Copeia, 1986*, 65–69.

Christian, K., Lawrence, W. T., & Snell, H. L. (1991). Effect of soil moisture on yolk and fat distribution in hatchlings lizards from natural nests. *Comparative Biochemistry and Physiology, 9*, 13–19.

Cochran, D. M. (1924). Notes on the herpetological collections made by Dr. W. L. Abbott on the Island of Haiti. *Proceedings of the United States National Museum, 66*, 1–15.

Cochran, D. M. (1931). New Bahamian reptiles. *Journal of the Washington Academy of Sciences, 21*, 39–40.

Coenen, C. (1995). Observations on the Bahamian rock iguana of the Exumas. *Bahamas Journal of Science, 2*, 8–14.

Cope, E. D. (1861). On an iguana from Andros Island. *Proceedings of the National Science Academy of Philadelphia, 13*, 123.

Cope, E. D. (1885). The large iguanas of the Greater Antilles. *American Naturalist, 19*, 1005–1006.

Cope, E. D. (1886). On the species of Iguaninae. *Proceedings of the American Philosophical Society, 23*, 261–271.

Coy Otero, A., & Lorenzo Hernandez, N. (1982). Lista de los helmintos parásitos de los vertebrados silvestres cubanos. *Serie Poeyana, 235*, 1–57.

Crissey, S. D., Ange, K. D., Jacobsen, K. L., Slifka, K. A., Bowen, P. E., Stacewicz-Sapuntizakis, M., Langman, C. B., Sadler, W., Kahn, S., & Ward, A. (2003). Serum concentrations of lipids, vitamin D metabolites, retinol, retinyl esters, tocopherols and selected carotenoids in twelve captive wild felid species at four zoos. *Journal of Nutrition, 133*, 160–166.

Crissey, S. D., Maslanka, M. M., and Ullrey, D. E. (1999). Assessment of nutritional status of captive and free ranging animals. Nutrition Advisory Group Handbook Fact Sheet 008. www.nagonline.net.

Cuvier, G. J. L. N. F. D. (1829). *Le Regne Animal Distribué, d'apres son Organisation, pur servir de base à l'Histoire naturelle des Animaux et d'introduction à l'Anatomie Comparé. Nouvelle Edition. Vol. 2: Les Reptiles*. Déterville, Paris.

Cyril, S., Hayes, W. K., & Carter, R. L. (2001). Taxon report. San Salvador iguana (*Cyclura rileyi rileyi*). *Newsletter of the IUCN Iguana Specialist Group, 4*(2), 5.

Divers, S. J. (2000). Reptilian renal and reproductive disease diagnosis. In A. M. Fudge (Ed.), *Laboratory Medicine, Avian and Exotic Pets* (pp. 217–222). Philadelphia, Pennsylvania: W.B. Saunders.

Divers, S. J., & Innis, C. J. (2006). Renal disease in reptiles: Diagnosis and clinical management. In D. R. Mader (Ed.), *Reptile Medicine and Surgery* (pp. 878–892). St Louis, MO: Elsevier Saunders.

Donoghue, S. (1995). Growth of juvenile green iguanas (*Iguana iguana*) fed four diets. *Journal of Nutrition, 124*, 2626S–2629S.

Donoghue, S. (2006). Nutrition. In D. Mader (Ed.), *Reptile Medicine and Surgery* (pp. 251–298). St Louis, MO: Elsevier Saunders.

Duméril, A. M. C., & Bibron, G. (1837). *Erpétologie Générale ou Histoire Naturelle Complete des Reptiles, Vol. 4*. Paris: Librairie Encyclopédique Roret.

Ehrig, R. (2000). West Indian iguana habitat. In A. C. Alberts (Ed.), *West Indian Iguanas: Status Survey and Conservation Action Plan* (p. 9). Gland, Switzerland: IUCN – the World Conservation Union.

Etheridge, R. (1964). Late Pleistocene lizards from Barbuda, British West Indies. *Bulletin of the Florida Academy of Sciences, 9*, 43–75.

Etheridge, R. (1966). Pleistocene lizards from New Providence. *Quarterly Journal of the Florida Academy of Sciences, 28*, 349–358.

Ferguson, G. W., Gehrmann, W. H., Peavy, B., Painter, C., Hartdegen, R., Chen, T. C., Holick, M. F., & Pinder, J. E. (2009). Restoring vitamin D in monitor lizards: Exploring the efficacy of dietary and UVB sources. *Journal of Herpetological Medicine and Surgery, 19*, 81–88.

Fisse, A., Draud, M., Raphael, R. L., & Melkonian, K. (2004). Differential leukocyte counts of critically endangered Grand Cayman Blue iguanas, *Cyclura nubila lewisi*. *Journal of Herpetological Medicine and Surgery, 14*, 4–19.

Fitzinger, L. J. F. J. (1843). *Systema Reptilium. Fasciculus Primus*. Vienna: Braumüller et Seidel.

Gallagher, R. P., & Lee, T. K. (2006). Adverse effects of ultraviolet radiation: a brief review. *Progress in Biophysics and Molecular Biology, 92*, 119–131.

Garcia, M. A., Perez-Buitrago, N., Alvarez, A. O., & Tolson, P. J. (2007). Survival, dispersal and reproduction of headstarted Mona Island iguanas. *Cyclura cornuta stejnegeri. Applied Herpetology, 4*, 357–363.

Gehrmann, W. H., Horner, J. D., Ferguson, G. W., Chen, T. C., & Holick, M. F. (2004). A comparison of responses by three broadband radiometers to different ultraviolet-B sources. *Zoo Biology, 23*, 355–363.

Gerber, G. P. (2000a). Lesser Caymans iguana: *Cyclura nubila caymanensis*. In A. C. Alberts (Ed.), *West Indian Iguanas: Status Survey and Conservation Action Plan* (pp. 41–43). Gland, Switzerland: IUCN – the World Conservation Union.

Gerber, G. P. (2000b). Conservation of the Anegada Iguana (*Cyclura pinguis*). Field research report to the British Virgin Islands National Parks Trust, Fauna and Flora International, and the Zoological Society of San Diego.

Gerber, G. P. (2007). Turks and Caicos iguana translocation program, Bahama Archipelago. *Re-introduction News, 26*, 53–55.

Gerber, G. P., & Iverson, J. B. (2000). Turks and Caicos iguana: *Cyclura carinata carinata*. In A. C. Alberts (Ed.), *West Indian Iguanas: Status Survey and Conservation Action Plan* (pp. 15–18). Gland, Switzerland: IUCN – the World Conservation Union.

Gerber, G. P., Alberts, A. C., Kearns, K. S., Keener, L., Czekala, N., & Morris, P. J. (2006). Veterinary contributions to the Turks and Caicos Iguana (*Cyclura carinata*) Restoration Project. *Proceedings of the American Association of Zoo Veterinarians, 2006*, 43.

Gerber, G. P., Grant, T. D., & Alberts, A. C. (2002). Lacertilia: *Cyclura nubila nubila* (*Cuban iguana*). Carrion feeding. *Herpetological Review, 33*, 133–134.

Gerber, G. P., Keener, L., Jezier, B., Czekala, N., MacDonald, E., & Alberts, A. C. (2004). Effects of translocation on the blood chemistry, hematology, and endocrinology of the critically endangered Turks and Caicos iguanas (*Cyclura carinata*). *Proceedings of the American Society of Zoo Veterinarians, 2004*, 572.

Gillespie, D., Frye, F., Stockham, S. L., & Fredeking, T. M. (2000). Blood values in wild and captive Komodo dragons. *Varanus komodoensis. Zoo Biology, 19*, 495–509.

Glor, R. E., Powell, R., & Parnerlee, J. S., Jr. (1998). *Cyclura ricordii. Catalogue of American Amphibians and Reptiles, 657*, 1–3.

Goodman, R. M. (2007). Activity patterns and foraging behavior of the endangered Grand Cayman blue iguana. *Cyclura lewisi. Caribbean Journal of Science, 43*, 73–86.

Goodman, R. M., & Burton, F. J. (2005). *Cyclura lewisi* hatchlings. *Herpetological Review, 36*, 176.

Goodman, R. M., Burton, F. J., & Echternacht, A. C. (2005). Habitat use of the endangered iguana *Cyclura lewisi* in a human-modified landscape on Grand Cayman. *Animal Conservation, 8*, 397–405.

Goodman, R. M., Knapp, C. R., Bradley, K. A., Gerber, G. P., & Alberts, A. C. (2009). Review of radio transmitter attachment methods for West Indian rock iguanas (genus *Cyclura*). *Applied Herpetology, 6*, 151–170.

Goodyear, N. C., & Lazell, J. D. (1994). Status of a relocated population of *Iguana pinguis* on Guana Island, British Virgin Islands. *Restoration Ecology, 2*, 43–50.

Gosse, P. H. (1848). On the habits of *Cyclura lophoma*, an iguanaform lizard. *Proceedings of the Zoological Society of London, 16*, 99–104.

Grant, C. (1940). The herpetology of the Cayman Islands. *Bulletin of the Institute of Jamaica Science Series, 2*, 1–65.

Gray, J. E. (1831). A synopsis of the species of the Class Reptilia. In G. Cuvier, E. Griffith, C. H. Smith, E. Pidgeon, J. E. Gray, & G. R. Gray (Eds.), *The Animal Kingdom* (pp. 1–481). London: Whittaker, Treacher, and Company.

Gray, J. E. (1845). *Catalogue of the Specimens of Lizards in the Collection of the British Museum*. London: Trustees of the British Museum/Edward Newman.

Gross, T. S., Guillette, L. J., Gross, D. A., & Cox, C. (1992). Control of oviposition in reptiles and amphibians. *Proceedings of the American Association of Zoo Veterinarians, 1992*, 158–166.

Harlan, J. (1824). Description of two new species of Linnaean *Lacerta* not before described, and construction of the new genus *Cyclura*. *Journal of the Academy of National Science of Philadelphia, 4*, 242–251.

Haq, B. U., Hardenbol, J., & Vail, P. R. (1987). Chronology of fluctuating sea levels since the Triassic. *Science, 235*, 1156–1166.

Hartley, L. M., Glor, R. E., Sproston, A. L., Powell, R., & Parmerlee, J. S., Jr. (2000). Germination rates of seeds consumed by two species of rock iguanas (*Cyclura* spp.) in the Dominican Republic. *Caribbean Journal of Science, 36*, 149–151.

Hayes, W. K. (2000a). White cay iguana: *Cyclura rileyi cristata*. In A. C. Alberts (Ed.), *West Indian Iguanas: Status Survey and Conservation Action Plan* (pp. 59–60). Gland, Switzerland: IUCN – the World Conservation Union.

Hayes, W. K. (2000b). San Salvador iguana: *Cyclura rileyi rileyi*. In A. C. Alberts (Ed.), *West Indian Iguanas: Status Survey and Conservation Action Plan* (pp. 56–58). Gland, Switzerland: IUCN – the World Conservation Union.

Hayes, W. K., & Montanucci, R. (2000). Acklins iguana: *Cyclura rileyi nuchalis*. In A. C. Alberts (Ed.), *West Indian Iguanas: Status Survey and Conservation Action Plan* (pp. 60–61). Gland, Switzerland: IUCN – the World Conservation Union.

Hayes, W. K., Carter, R. L., Cyril, S., Jr., & Thornton, B. (2004). Conservation of an endangered Bahamian rock iguana. I. In A. C. Alberts, R. L. Carter, W. K. Hayes, & E. P. Martins (Eds.), *Iguanas: Biology and Conservation* (pp. 232–257). Los Angeles: University of California Press.

Hazard, L. C. (2004). Sodium and potassium secretion by iguana salt glands. In A. C. Alberts, R. L. Carter, W. K. Hayes, & E. P. Martins (Eds.), *Iguanas: Biology and Conservation* (pp. 894–913). Los Angeles: University of California Press.

Hedges, S. B. (2001). Biogeography of the West Indies: An overview. In C. A. Woods, & F. E. Sergile (Eds.), *Biogeography of the West Indies: Patterns and Perspectives* (2nd edn.). (pp. 15–33) Boca Raton, FL: CRC Press.

Hines, K. (2007). Preliminary diet analyses for *Cyclura cychlura inornata* and *Cyclura cychlura figginsi*: Assessing potential impacts of tourist feeding. *Newsletter of the IUCN Iguana Specialist Group, 10*(1), 13.

Hines, K., & Iverson, J. B. (2006). Allen Cays iguana. *Newsletter of the IUCN Iguana Specialist Group, 9*(1), 10–11.

Holick, M. F. (1990). The use and interpretation of assays for vitamin D and its metabolites. *Journal of Nutrition, 120*, 1464–1469.

Holick, M. F. (1995). Environmental factors that influence the cutaneous production of vitamin D. *American Journal of Clinical Nutrition, 61*, S638–S645.

Holick, M. F., Tian, X. Q., & Allen, M. E. (1995). Evolutionary importance for the membrane enhancement of the production of vitamin D3 in the skin of poikilothermic animals. *Proceedings of the National Academy of Sciences, 92*, 3124–3126.

Hudson, R. D., & Alberts, A. C. (2004). The role of zoos in the conservation of West Indian iguanas. In A. C. Alberts, R. L. Carter, W. K. Hayes, & E. P. Martins (Eds.), *Iguanas: Biology and Conservation* (pp. 274–289). Los Angeles: University of California Press.

Iturralde-Vinent, M. A. (2006). Meso-Cenozoic Caribbean paleogeography: Implications for the historical biogeography of the region. *International Geology Review, 48*, 791–827.

IUCN Iguana Specialist Group. (2008). Andros Iguana *(Cyclura cychlura cychlura)* Conservation Action Plan 2009-2011. Gland, Switzerland: IUCN – the World Conservation Union.

Iverson, J. B. (1978). The impact of feral cats and dogs on populations of the West Indian rock iguana. *Cyclura carinata. Biological Conservation, 14*, 63–73.

Iverson, J. B. (1979). Behavior and ecology of the rock iguana. *Cyclura carinata. Bulletin of the Florida State Museum of Biological Sciences, 24*, 175–358.

Iverson, J. B. (1982). Adaptations to herbivory in Iguanine lizards. In G. M. Burghardt, & A. S. Rand (Eds.), *Iguanas of the World* (pp. 60–76). Park Ridge, NJ: Noyes Publications.

Iverson, J. B. (1985). Lizards as seed dispersers? *Journal of Herpetology, 19*, 292–293.

Iverson, J. B. (1989). Natural growth in the Bahamian iguana. *Cyclura cychlura. Copeia, 1989*, 502–505.

Iverson, J. B. (2000). Allen's Cay iguana: *Cyclura cychlura inornata*. In A. C. Alberts (Ed.), *West Indian Iguanas: Status Survey and Conservation Action Plan* (pp. 34–36). Gland, Switzerland: IUCN – the World Conservation Union.

Iverson, J. B. (2007). Juvenile survival in the Allen Cays rock iguana (*Cyclura cychlura inornata*). *Copeia, 2007*, 740–744.

Iverson, J. B., & Mamula, M. R. (1989). Natural growth in the Bahamian iguana. *Cyclura cychlura. Copeia, 1989*, 502–505.

Iverson, J. B., Converse, S. J., Smith, G. R., & Valiulis, J. M. (2006b). Long-term trends in the demography of the Allen Cays rock iguana (*Cyclura cychlura inornata*): Human disturbance and density-dependent effects. *Biological Conservation, 132*, 300–310.

Iverson, J. B., Hines, K. N., & Valiulis, J. M. (2004). The nesting ecology of the Allen Cays rock iguana, *Cyclura cychlura inornata*, in the Bahamas. *Herpetological Monographs, 18*, 1–36.

Iverson, J. B., Pasachnik, S. A., Knapp, C. R., & Buckner, S. D. (2006a). *Cyclura cychlura. Catalogue of American Amphibians and Reptiles, 810*, 1–8.

Jacobson, E. R. (2003). *Biology, Husbandry, and Medicine of the Green Iguana*. Malabar, FL: Kreiger.

James, S. B., Iverson, J., Greco, V., & Raphael, B. L. (2006). Health assessment of Allen Cays rock iguana, *Cyclura cychlura inornata. Journal of Herpetological Medicine and Surgery, 16*, 93–98.

Jordan, M. J. R. (2005). Dietary analysis for mammals and birds: A review of field techniques and animal-management applications. *International Zoo Yearbook, 39*, 108–116.

Juan-Salles, C., Garner, M. M., Monreal, T., & Burgos-Rodriguez, A. G. (2008). Ovarian torsion in a green, *Iguana iguana*, and a Rhinoceros, *Cyclura cornuta*, iguana. *Journal of Herpetological Medicine and Surgery, 18*, 14–17.

Klasing, K. C. (1998). Comparative Avian Nutrition. *CAB International*. Cambridge: University Press.

Knapp, C. R. (2000a). Exuma Island iguana: *Cyclura cychlura figginsi*. In A. C. Alberts (Ed.), *West Indian Iguanas: Status Survey and Conservation Action Plan* (pp. 32–34). Gland, Switzerland: IUCN – the World Conservation Union.

Knapp, C. R. (2000b). Home range and intraspecific interactions of a translocated iguana population (*Cyclura cychlura inornata* Barbour and Noble). *Caribbean Journal of Science, 36*, 250–257.

Knapp, C. R. (2001). Status of a translocated *Cyclura iguana* colony in the Bahamas. *Journal of Herpetology, 35*, 239–248.

Knapp, C. R. (2004). Ecotourism and its impact on iguana conservation. In A. C. Alberts, R. L. Carter, W. K. Hayes, & E. P. Martins (Eds.), *Iguanas: Biology and Conservation* (pp. 290–301). Los Angeles: University of California Press.

Knapp, C. R. (2005a). *Ecology and conservation of the Andros iguana* (Cyclura cychlura cychlura). PhD Dissertation. Gainesville, FL: University of Florida.

Knapp, C. R. (2005b). Working to save the Andros iguana. *Iguana, 12*, 9–13.

Knapp, C. R., & Hudson, R. D. (2004). Translocation strategies as a conservation tool for West Indian iguanas: Evaluations and recommendations. In A. C. Alberts, R. L. Carter, W. K. Hayes, & E. P. Martins (Eds.), *Iguanas: Biology and Conservation* (pp. 199–209). Los Angeles: University of California Press.

Knapp, C. R., & Malone, C. L. (2003). Patterns of reproductive success and genetic variability in a translocated iguana population. *Herpetologica, 59*, 195–202.

Knapp, C. R., & Owens, A. K. (2004). Diurnal refugia and novel ecological attributes of the Bahamian boa *Epicrates striatus fowleri* (Boidae). *Caribbean Journal of Science, 40*, 265–270.

Knapp, C. R., & Owens, A. K. (2005). Home range and habitat associations of a Bahamian iguana: implications for conservation. *Animal Conservation, 8*, 269–278.

Knapp, C. R., Alvarez-Clare, S., & Perez-Heydrich, C. (2010). The influence of landscape heterogeneity and dispersal on survival of neonate insular iguanas. *Copeia, 2010*, 62–70.

Knapp, C. R., Iverson, J. B., & Owens, A. K. (2006). Geographic variation in nesting behavior and reproductive biology of an insular iguana (*Cyclura cychlura*). *Canadian Journal of Zoology, 84*, 1566–1575.

Lacepède, B. G. E. (1789). *Histoire Naturelle des Serpens*. Imprimerie du Roi (Tome Second). Paris: Hôtelde Thou. (1 unnumbered page) + 1-8 + 1-19, Errata, 1-144 + 1-527.

Lacy, K. E., & Martins, E. P. (2003). The effect of anthropogenic habitat usage on the social behaviour of a vulnerable species,. *Cyclura nubila. Animal Conservation, 6*, 3–9.

Lemm, J., Steward, S., & Schmidt, T. (2005). Captive reproduction of the critically endangered Anegada Island iguana (*Cyclura pinguis*) at the San Diego Zoo. *International Zoo Yearbook, 39*, 141–152.

Lewis, C. B. (1944). Notes on *Cyclura*. *Herpetologica, 2*, 93–98.

Lewis, D.S., van Veen, R., and Wilson, B. S. (2011). Conservation implications of small Indian mongoose (*Herpestes auropunctatus*) predation in a hotspot within a hotspot: the Hellshire Hills, Jamaica. *Biological Invasions, 13*, 25–33.

Lewis, J. F., & Draper, G. (1990). Geology and tectonic evolution of the northern Caribbean margin. In G. Dengo, & J. E. Case (Eds.), *The Caribbean Region* (pp. 77–140). Boulder, CO: Geological Society of America.

Laing, C. J., & Fraser, D. R. (1999). The vitamin D system in iguanian lizards. *Comparative Biochemistry and Physiology B, 123*, 373–379.

Laing, C. J., Trube, A., Shae, G. M., & Fraser, D. R. (2001). The requirements for natural sunlight to prevent vitamin D deficiency in iguanian lizards. *Journal of Zoo and Wildlife Medicine, 32*, 342–348.

Lloyd, M. L. (1990). Reptilian dystocias review: Causes, prevention, management, and comments on the synthetic hormone vasoticin. *Proceedings of the American Association of Zoo Veterinarians, 2002*, 290–296.

Lung, N. P., Raphael, B. L., Ramer, J. C., & Reichard, T. A. (2002). West Indian rock iguana conservation: The importance of veterinary involvement. *Proceedings of the American Association of Zoo Veterinarians, 2002*, 234.

MacPhee, R. D. E., Iturralde-Vinent, M. A., & Gaffney, E. S. (2003). Domo de Zaza, an early Miocene vertebrate locality in south-central Cuba, with notes on the tectonic evolution of Puerto Rico and the Mona passage. *American Museum of Natural History Novitates, 3394*, 1–42.

Mader, D. R. (2006). *Reptile Medicine and Surgery* (2nd edn.). St Louis, MO: Elsevier Saunders.

Malone, C. L., & Davis, S. K. (2004). Genetic contributions to Caribbean iguana conservation. In A. C. Alberts, R. L. Carter, W. K. Hayes, & E. P. Martins (Eds.), *Iguanas: Biology and Conservation* (pp. 45–57). Los Angeles: University of California Press.

Malone, C. L., Knapp, C. R., Taylor, J. F., & Davis, S. K. (2003). Genetic consequences of Pleistocene fragmentation: Isolation, drift, and loss of diversity in rock iguanas (*Cyclura*). *Conservation Genetics, 4*, 1–15.

Malone, C. L., Wheeler, T., Taylor, J. F., & Davis, S. K. (2000). Phylogeography of the Caribbean rock iguana (*Cyclura*): Implications for conservation and insights on the biogeographic history of the West Indies. *Molecular Phylogenetics and Evolution, 17*, 269–279.

Maria, R., Ramer, J. C., Reichard, T. A., Tolson, P. J., & Christopher, M. M. (2007). Biochemical reference intervals and intestinal microflora of free-ranging Ricord's iguanas (*Cyclura ricordii*). *Journal of Zoo and Wildlife Medicine, 38*, 414–419.

Martins, E. P., & Lacy, K. E. (2004). Behavior and ecology of rock iguanas I: Evidence for an appeasement display. In A. C. Alberts, R. L. Carter, W. K. Hayes, & E. P. Martins (Eds.), *Iguanas: Biology and Conservation* (pp. 101–108). Los Angeles: University of California Press.

Martins, E. P., & Lamont, J. (1998). Evolution of communication and social behavior: a comparative study of *Cyclura* rock iguanas. *Animal Behaviour, 55*, 1685–1706.

McBee, R. H., & McBee, V. H. (1982). The hindgut fermentation in the green iguana, *Iguana iguana*. In G. M. Burghardt, & A. S. Rand (Eds.), *Iguanas of the World* (pp. 77–83). Park Ridge, NJ: Noyes Publications.

Miller, G. S. (1918). Mammals and reptiles collected by Theodore de Booy in the Virgin Islands. *Proceedings of the US National Museum, 54*, 507–511.

Mitchell, N. C. (1999). Effect of introduced ungulates on density, dietary preferences, home range, and physical condition of the iguana *Cyclura pinguis on Anegada. Herpetologica, 55*, 7–17.

Mitchell, N. C. (2000). Anegada Island iguana: *Cyclura pinguis*. In A. C. Alberts (Ed.), *West Indian Iguanas: Status Survey and Conservation Action Plan* (pp. 47–49). Gland, Switzerland: IUCN – the World Conservation Union.

Murphy, P. G., & Lugo, A. E. (1986). Ecology of dry tropical forest. *Annual Review of Ecology and Systematics, 17*, 67–88.

Nagy, K. A. (1982). Energy requirements of free-living Iguanid lizards. In G. M. Burghardt, & A. S. Rand (Eds.), *Iguanas of the World* (pp. 49–59). Park Ridge, NJ: Noyes Publications.

National Research Council. (1977). *Nutrient Requirements of Rabbits* (2nd rev. edn.). Washington, DC: National Academy of Sciences.

National Research Council. (1994). *Nutrient Requirements of Poultry* (9th rev. edn.). Washington, DC: National Academy of Sciences.

National Research Council. (2006). *Nutrient Requirements of Dogs and Cats*. Washington, DC: National Academy of Sciences.

National Research Council. (2007). *Nutrient Requirements of Horses* (6th rev. edn.). Washington, DC: National Academy of Sciences.

Ottenwalder, J. (2000a). Rhinoceros iguana: *Cyclura cornuta cornuta*. In A. C. Alberts (Ed.), *West Indian Iguanas: Status Survey and Conservation Action Plan* (pp. 22–27). Gland, Switzerland: IUCN – the World Conservation Union.

Ottenwalder, J. (2000b). Ricord's iguana: *Cyclura ricordii*. In A. C. Alberts (Ed.), *West Indian Iguanas: Status Survey and Conservation Action Plan* (pp. 51–55). Gland, Switzerland: IUCN – the World Conservation Union.

Ottenwalder, J. (2000c). Socioeconomic perspective. In A. C. Alberts (Ed.), *West Indian Iguanas: Status Survey and Conservation Action Plan* (pp. 3–7). Gland, Switzerland: IUCN – the World Conservation Union.

Perera, A. (2000). Cuban iguana: *Cyclura nubila nubila*. In A. C. Alberts (Ed.), *West Indian Iguanas: Status Survey and Conservation Action Plan* (pp. 36–40). Gland, Switzerland: IUCN/SSC IUCN – the World Conservation Union.

Perez-Buitrago, N., & Sabat, A. (2000). Population status of the rock ground iguana (*Cyclura cornuta stejnegeri*) in Mona Island, Puerto Rico. *Acta Cientifica, 14*, 67–76.

Perez-Buitrago, N., & Sabat, A. (2007). Natal dispersal, home range, and habitat use of hatchlings of the Mona Island iguana (*Cyclura cornuta stejnegeri*). *Applied Herpetology, 4*, 365–376.

Perez-Buitrago, N., Alvarez, A. O., & Garcia, M. A. (2006). Cannibalism in an introduced population of *Cyclura nubila nubila* on Isla Magueyes, Puerto Rico. *Iguana, 13*, 206–208.

Perez-Buitrago, N., Garcia, M. A., Sabat, A., Delgado, J., Alvarez, A., McMillan, O., & Funk, S. M. (2008). Do headstart programs work? Survival and body condition in headstarted Mona Island iguanas. *Cyclura cornuta stejnegeri. Endangered Species Research, 6*, 55–65.

Perez-Buitrago, N., Sabat, A., Funk, S. M., Garcia, M. A., Alvarez, A. O., & McMillan, W. O. (2007). Spatial ecology of the Mona Island iguana *Cyclura cornuta stejnegeri* in an undisturbed environment. *Applied Herpetology, 4*, 347–355.

Perry, G., & Garland, T., Jr. (2002). Lizard home ranges revisited: Effects of size, body size, diet, habitat, and phylogeny. *Ecology, 83*, 1870–1885.

Perry, G., Lazell, J., LeVering, K., & Mitchell, N. (2007). Body size and timing of reproduction in the highly endangered stout iguana, *Cyclura pinguis*, in the British Virgin Islands. *Caribbean Journal of Science, 43*, 155–159.

Pindell, J., Kennan, L., Stanek, K. P., Maresch, W. V., and Draper, G. (2006). Foundations of Gulf of Mexico and Caribbean evolution: Eight controversies resolved. In M. A. Iturralde-Vinent, & E. G. Lidiak (Eds.), *Caribbean Plate Tectonics: Stratigraphic, Magmatic, Metamorphic, and Tectonic Events.* UNESCO/IUGS IGCP Project 433. *Geologica Acta* 4(1–2), pp. 303–341.

Plumb, D. C. (2008). *Plumb's Veterinary Drug Handbook* (6th edn.). Ames, IO: Blackwell.

Powell, R. (1999). Herpetology of Navassa Island, West Indies. *Caribbean Journal of Science, 35*, 1–13.

Powell, R. (2000a). Horned iguanas of the Caribbean. *Reptile and Amphibian Hobbyist, 5*(12), 30–37.

Powell, R. (2000b). *Cyclura onchiopsis. Catalogue of American Amphibians and Reptiles, 710*, 1–3.

Powell, R., & Glor, R. E. (2000). *Cyclura stejnegeri. Catalogue of American Amphibians and Reptiles, 711*, 1–4.

Pregill, G. (1981). Late Pleistocene herpetofauna from Puerto Rico. University of Kansas Museum of Natural History, Miscellaneous Publication 71: 1–72.

Pregill, G. (1982). Fossil amphibians and reptiles from New Providence Island, Bahamas. In S. L. Olson (Ed.), *Fossil Vertebrates from the Bahamas* (pp. 8–21). Washington, DC: Smithsonian Scholarly Press, Smithsonian Contributions to Paleobiology 48.

Raila, J., Schuhmacher, A., Gropp, J., & Schweigert, F. J. (2002). Selective absorption of carotenoids in the common green iguana (*Iguana iguana*). *Comparative Biochemistry and Physiology A, 132*, 513–518.

Ramer, J. C., Maria, R., & Reichard, T. A. (2005). Vitamin D status of wild Ricord's iguanas (*Cyclura ricordii*) and captive and wild Rhinoceros iguanas (*Cyclura cornuta cornuta*) in the Dominican Republic. *Journal of Zoo and Wildlife Medicine, 36*, 188–191.

Ramer, J. C., Williams, J., Roedell, L., Accime, M., & Rupp, E. (2009). Comprehensive conservation program for Ricord's iguana (*Cyclura ricordii*) in the Dominican Republic and Haiti. *Proceedings of the American Association of Zoo Veterinarians, 2009*, 76.

Raphael, B. L. (2004). The Iguana Specialist Group veterinary update. *Proceedings of the American Association of Zoo Veterinarians, 2004*, 329.

Raphael, B. L. (2006). An update on rock iguana (*Cyclura* spp.) headstart projects. *Proceedings of the American Association of Reptile and Amphibian Veterinarians, 2006*, 32.

Richman, L. K., Montali, R. K., Allen, M. E., & Oftedal, O. T. (1995). Paradoxical pathological changes in vitamin D deficient green iguanas (*Iguana iguana*). *Proceedings of the American Association of Zoo Veterinarians, 1995*, 231–232.

Rivero, J. A. (1978). *Los anfibios y reptiles de Puerto Rico*. Rio Piedras, Puerto Rico: Editorial Universitaria, Universidad de Puerto Rico.

Rupp, E., Inchaustegui, S., & Arias, Y. (2005). Conservation of *Cyclura ricordii* in the Southwestern Dominican Republic and a brief history of the Grupo Jaragua. *Iguana, 12*, 223–233.

Rupp, E., Inchaustegui, S., & Arias, Y. (2008). Conserving *Cyclura ricordii* 2007. *Iguana, 15*, 2–7.

Scheelings, F. (2008). Pre-ovulatory follicular stasis in a yellow-spotted monitor, *Varanus panoptes panoptes*. *Journal of Herpetological Medicine and Surgery, 18*, 18–20.

Schmidt, K. P. (1920). Some new and rare amphibians and reptiles from Cuba. *Proceedings of the Linnaean Society of New York, 33*, 3–8.

Schmidt, D. A., Kerley, M. S., Dempsey, J. L., & Porton, I. J. (1999). The potential to increase neutral detergent fiber levels in ape diets using readily available produce. *Proceedings of the 3rd Biannual Association of Zoos and Aquariums Nutrition Advisory Group Conference on Zoo and Wildlife Nutrition, 1999*, 102–107.

Schmidt, D. A., Mulkerin, D., Boehm, D. R., Lu, Z., Ellersieck, M. R., Campbell, M., Chen, T. C., & Holick, M. F. (2006). *Proceedings of the 6th Comparative Nutrition Society Symposium, 2006*, 162–166.

Schwartz, A., & Carey, M. (1977). Systematics and evolution in the West Indian iguanid genus *Cyclura*. *Studies on the Fauna of Caraçao and Other Caribbean islands, 53,* 16–97.

Schwartz, A., & Thomas, R. (1975). A checklist of West Indian amphibians and reptiles. *Carnegie Museum of Natural History Special Publication, 1,* 1–216.

Schwartz, A., & Henderson, R. W. (1991). *Amphibians and Reptiles of the West Indies: Descriptions, Distributions, and Natural History.* Gainesville, FL: University of Florida Press.

Shrestha, S. P., Ullrey, D. E., Bernard, J. B., Wemmer, C., & Kraemer, D. C. (1998). Plasma vitamin D and other analyte levels in Nepalese camp elephants (*Elephas maximus*). *Journal of Zoo and Wildlife Medicine, 29,* 260–278.

Slavens, F. L., & Slavens, K. (1994). *Reptiles and Amphibians in Captivity: Breeding, Longevity, and Inventory.* Seattle: Slaveware.

Slifka, K. A., Bowen, P. E., Stacewicz-Sapuntzakis, M., & Crissey, S. D. (1999). A survey of serum and dietary carotenoids in captive wild animals. *Journal of Nutrition, 129,* 380–390.

Smith, G. R., & Iverson, J. B. (2006). Changes in sex ratio over time in the endangered iguana. *Cyclura cychlura inornata. Canadian Journal of Zoology, 84,* 1522–1527.

Snell, H. L., & Christian, K. A. (1985). Energetics of Galapagos Land iguanas: A comparison of two island populations. *Herpetologica, 41,* 437–442.

Stejneger, L. (1903). A new species of large iguana from the Bahama Islands. *Proceedings of the United States National Museum, 26,* 129–132.

Telford, S. R., Jr. (1984). Hemoparasites of reptiles. In G. L. Hoff, F. L. Frye, & E. R. Jacobson (Eds.), *Diseases of Amphibians and Reptiles* (pp. 385–517). New York: Plenum Press.

Thornton, B. J. (2000). *Nesting ecology of the endangered Acklins Bight rock iguana,* Cyclura rileyi nuchalis, *in the Bahamas.* Master's thesis. Andrews University.

Tolson, P. J. (2000). Control of introduced species. In A. C. Alberts (Ed.), *West Indian Iguanas: Status Survey and Conservation Action Plan* (pp. 86–89). Gland, Switzerland: IUCN – the World Conservation Union.

Troyer, K. (1983). Review: The biology of Iguanine Lizards: present status and future directions. *Herpetologica, 39,* 317–328.

Ullrey, D. E., & Bernard, J. B. (1999). Vitamin D: Metabolism, sources, unique problems in zoo animals, meeting needs. In M. E. Fowler, & R. E. Miller (Eds.), *Zoo and Wild Animal Medicine* (4th edn.). (pp. 63–78). Philadelphia, PA: W.B. Saunders.

van der Wardt, S. T., Kik, M. J. L., Klaver, P. S. J., Janse, M., & Beynen, A. C. (1999). Calcium balance in Drakensberg crag lizards (*Pseudocordylus melanotus melanotus: Cordylidae*). *Journal of Zoo and Wildlife Medicine, 30,* 541–544.

van Marken Lichtenbelt, W. D., Vogel, J. T., & Wesselingh, R. A. (1997). Energetic consequences of field body temperatures in the green iguana. *Ecology, 78,* 297–307.

Vogel, P. (1994). Evidence of reproduction in a remnant population of the endangered Jamaican iguana, *Cyclura collei* (Lacertilia, Iguanidae). *Caribbean Journal of Science, 30,* 234–241.

Vogel, P. (2000). Jamaican iguana: *Cyclura collei.* In A. C. Alberts (Ed.), *West Indian Iguanas: Status Survey and Conservation Action Plan* (pp. 19–22). Gland, Switzerland: IUCN – the World Conservation Union.

Vogel, P., Nelson, R., & Kerr, R. (1996). Conservation strategy for the Jamaican iguana, *Cyclura collei.* In R. Powell., & R. W. Henderson. (Eds.), *Contributions to West Indian Herpetology: A Tribute to Albert Schwartz* (pp. 395–406). Ithaca, NY: Society for the Study of Amphibians and Reptiles.

Wagler, J. G. (1830). *Naturliches System der Amphibien mit vorangehender Classification der Saugethiere und Vogel.* Munich: J.G. Cott'schen Buchhandlung. Stuttgart and Tubingen.

Wasilewski, J., & Conners, S. (2005). Bartsch's rock iguana (*Cyclura carinata bartschi*) update. *Iguana Specialist Group Newsletter, 8*(2), 13.

Welch, M. E., Gerber, G. P., & Davis, S. K. (2004). Genetic structure of the Turks and Caicos Rock Iguana and its implications for species conservation. In A. C. Alberts, R. L. Carter, W. K. Hayes, & E. P. Martins (Eds.), *Iguanas: Biology and Conservation* (pp. 58–70). Los Angeles: University of California Press.

Wiewandt, T. A. (1977). *Ecology, behavior, and management of the Mona Island ground iguana,* Cyclura stejnegeri. PhD dissertation. Ithaca, New York: Cornell University.

Wiewandt, T., & Garcia, M. (2000). Mona Island iguana: *Cyclura cornuta stejnegeri.* In A. C. Alberts (Ed.), *West Indian Iguanas: Status Survey and Conservation Action Plan* (pp. 27–31). Gland, Switzerland: IUCN – the World Conservation Union.

Wilcox, K., Carter, J. Y., & Wilcox, L. V., Jr. (1973). Range extension of *Cyclura figginsi* Barbour in the Bahamas. *Caribbean Journal of Science, 13,* 3–4.

Wilson, B. S., & van Veen, R. (2006). Update: Jamaican Iguana Recovery Project. *Newsletter of the IUCN Iguana Specialist Group, 8*(2), 4–6.

Wilson, B. S., & van Veen, R. (2007). Update: Jamaican Iguana Recovery Project. *Iguana Specialist Group Newsletter, 10*(1), 5–7.

Wilson, B. S., Alberts, A. C., Graham, K. S., Hudson, R. D., Bjorkland, R. K., Lewis, D. S., Lung, N. P., Nelson, R., Thompson, N., Kunna, J. L., & Vogel, P. (2004a). Survival and reproduction of repatriated Jamaican iguanas: Headstarting as a viable conservation strategy. In A. C. Alberts, R. L. Carter, W. K. Hayes, & E. P. Martins (Eds.), *Iguanas: Biology and Conservation* (pp. 220–231). Los Angeles: University of California Press.

Wilson, B. S., Robinson, O. F., & Vogel, P. (2004b). Status of the Jamaican iguana (*Cyclura collei*): Assessing 15 years of conservation effort. *Iguana, 11,* 225–231.

Windrow, S. L. (1977). *Winter activity and behavior of the Exuman rock iguana, Cyclura cychlura figginsi.* Master's thesis. New Brunswick, NJ: Rutgers University.

Woodley, J. D. (1980). Survival of the Jamaican iguana, *Cyclura collei. Journal of Herpetology, 14,* 45–49.

Zachariah, T. T., Knapp, C. R., Romero, L. M., & Singh, R. S. (2009). Conservation of the Andros iguana (*Cyclura cychlura cychlura*). *Proceedings of the American Association of Zoo Veterinarians, 2009,* 84.

Glossary

Aerobic Metabolic activities that rely on oxidation to provide energy.
Ambient temperature The temperature of the surrounding environment.
Anaerobic Metabolic energy obtained without the use of oxygen.
Anorexia Loss of appetite or refusal to feed.
Anterior Toward the front.
Arboreal Bush- or tree-dwelling.
Atrophy Wasting of tissue or an organ.
Bask To warm the body in sunlight or other heat source.
Caribbean Pertaining to the Caribbean Sea.
Caribbean Sea A sea of the Atlantic Ocean situated in the tropics of the Western hemisphere; it is bounded to the southwest by the Central American countries of Panama, to the west by Costa Rica, Nicaragua, Honduras, Guatemala, Belize, and Mexico, to the north by the Greater Antilles, and to the east by the Lesser Antilles.
Carrion The remains of a dead animal.
Caudal Pertaining to the tail.
Chemoreception Detection of chemicals.
Clade A group of descendant species that share a common ancestor.
Cloaca The common chamber into which the digestive, urinary, and reproductive systems empty and which opens exteriorly through the cloacal vent.
Conspecifics Members of the same species.
Dewlap A fan-like extension of the throat and/or lower neck.
Dimorphic A difference in form, build, or coloration involving the same species; often sex-linked.
Distal A part of an appendage that is away from the body.
Diurnal Active by day.
Dorsal Pertaining to the back or upper surface.
Dorsolateral Pertaining to the upper sides.
Dorsum The upper surface.
Ecdysis The process of molting the epidermis (skin).
Ectotherm An animal that relies on environmental heat sources to gain or lose heat.
Endemic A taxon restricted to a specific geographical region.
Extant Surviving to the present day.
Fecundity Number of eggs or offspring produced.
Femoral pores Glands found on the underside of the hind legs that produce lipid- and protein-based substances used to deposit scent trails.
Flora Referring to plants.
Fossorial Burrowing or underground lifestyle.
Genera Plural of genus.
Genus A taxonomic classification of a group of species having similar characteristics; the taxonomic designation below family and above species.
Gradient (temperature) A linear temperature change from low to high.
Gravid Referring to females containing eggs.
Greater Antilles The islands of Cuba, Jamaica, Hispaniola, and Puerto Rico.
Gular Pertaining to the throat.

Hemipenes The dual copulatory organs of male lizards and snakes.
Hemipenis The singular form of hemipenes.
Herbivorous Plant or vegetation eating.
Herpetology The study of amphibians and reptiles.
Hindgut fermentation In herbivores, the process of acquiring energy wherein microbes in the large intestine ferment vegetation and release volatile fatty acids as waste products.
Hybrid Offspring resulting from the breeding of two different species.
Hypervitaminosis Pertaining to a specific vitamin overdosage.
Hypovitaminosis Pertaining to a specific vitamin deficiency.
Insular Of, relating to, or constituting an island.
Interorbital The region between the eyes.
Interspace The patch of color between two markings.
Juvenile A young or immature specimen.
Keel A ridge; usually pertaining to the center of a scale or a row of scales.
Labial Pertaining to the lips.
Lateral Pertaining to the side.
Melanism A profusion of black pigment; dark or black specimens are often termed "melanistic."
Mental The scale at the tip of the lower lip.
Microflora Bacterial and protozoan inhabitants of the intestine.
Mid-dorsal Pertaining to the middle of the back.
Midventral Pertaining to the center of the abdomen or belly.
Nares Nostrils.
Necropsy Medical examination of a corpse.
Nocturnal Active at night.
Ocelli Eyelike spots.
Olfactory Referring to the sense of smell.
Omnivorous Eating both animal and plant material.
Ontogenetic Age-related changes.
Oviparous Reproducing by means of eggs that hatch after laying.
Pathogen A disease-causing organism.
Phenotype An organism's external appearance.
Pheromone A chemical used in communication.
Photoperiod Light—dark cycle.
Phylogenetics Determination of evolutionary relationships (phylogeny) through reconstruction or estimation of the pattern of ancestry and descent.
Plasma The fluid portion of unclotted blood.
Posterior Behind or to the rear.
Postocular Behind or to the rear of the eye.
Refugia Burrows, crevices, rock cracks, or other hiding places.
Renal Pertaining to the kidneys.
Rostral The scale on the tip of the snout.
Species A group of morphologically similar individuals that produce viable offspring; the taxonomic designation beneath genus and above subspecies.
Subcutaneous Referring to beneath the skin.
Subspecies The subdivision of a species. Subspecies often differ slightly from one another in color, size, scalation, or other criteria.
Snout—vent length (SVL) A linear measurement from the tip of the snout to the cloacal opening.
Sympatric Occurring together in the same geographic region.
Taxon A named taxonomic unit, such as species, genus, family, or higher unit of classification.

Taxonomy The science of classification of plants and animals.
Terrestrial Land-dwelling.
Thermoregulate The ability to regulate body temperature by choosing a warmer or cooler environment.
Umbilicus Navel.
Urates Urinary salts or uric acid, usually sodium, potassium, or ammonia.
Vent The external opening of the cloaca.
Venter The underside or belly of a specimen.
Ventral Pertaining to the underside or belly.
Ventrolateral Pertaining to the sides of the belly or venter.
Xeric Relatively arid.

Index

Acklin's iguana
  conservation status, 67–68
  description, 65
  natural history, 66–67
  population estimate, 176
  synonyms, 65
  taxonomic notes, 67
Action Plan for West Indian Iguanas, 199
Allen Cays iguana
  conservation status, 41
  description, 38–39
  natural history, 39–41
  population estimate, 176
  synonyms, 38
  taxonomic notes, 41
Amoebaiasis, management, 152
Andros Island iguana
  conservation status, 34
  description, 30–32
  evolution, 10–11
  habitat protection, 185–187
  natural history, 32–33
  population estimate, 176
  synonyms, 30
  taxonomic notes, 34
Anegada Island iguana
  conservation status, 59
  description, 54–55
  evolution, 5
  habitat protection, 185–187
  headstarting, 189–192
  natural history, 55–58
  phylogenetic tree, 9
  population estimate, 176
  synonyms, 54
  taxonomic notes, 59
Association of Zoos and Aquariums (AZA), 194–195, 197–198
Automobile, iguana fatalities, 185
AZA, *see* Association of Zoos and Aquariums

Bacterial infection, management, 153, 170
Bahamas, rock iguana evolution, 9
Biogeography, 5–9
Birds, iguana predation, 84, 86, 91
Bladder stones, *see* Cystic calculi
Blindfold, restraint for measuring, 112
Blood
  collection technique, 161–162
  health assessment and establishment of normal values, 168–170
  vitamin data from serum, 142–143
Body measurements, rock iguana, 17, 112, 125
Body posturing, communication, 88
Booby Cay, rock iguanas, 10, 19
Breeding, *see* Reproduction

Calcium
  deficiency and metabolic bone disease, 140–141, 158, 171
  dietary requirements, 134, 136
Cannibalism, hatchlings, 83
Capture, captive iguanas, 110
Caribbean, geological changes, 6
Cat
  control, 187–188
  iguana predation, 11, 84–85, 171, 182–183
Cloaca, swabbing, 165
Clutch size, 91
Colon, complexity, 80
Conservation
  habitat loss and degradation, 178
  habitat protection, 185–187
  headstarting, 189–192
  human interactions, 183–185
  hunting and poaching, 182–183
  invasive species control, 187–189
  long-term planning, 199
  organizations, 193–195
  outreach and education, 196–199
  population estimates, 175–177
  predation by introduced mammals, 11, 84–86, 180–182
  research, 195–196
  status by species
    Acklin's iguana, 67
    Allen Cays iguana, 41
    Andros Island iguana, 34
    Anegada Island iguana, 59
    Cuban iguana, 54
    Exuma Island iguana, 37

Conservation (*Continued*)
    Grand Cayman Blue iguana, 45
    Jamaican iguana, 24
    Mona Island iguana, 30
    Navassa Island iguana, 72
    rhinoceros iguana, 27
    Ricord's iguana, 62
    San Salvador iguana, 71
    Sister Isles iguana, 49
    Turks and Caicos iguana, 21
    White Cay iguana, 64
  translocation programs, 192–193
  zoos, 193–194
Courtship behavior, 89–90
Cryptosporidium, infection, 153
Cuban iguana
  breeding, 112
  conservation status, 54
  description, 50–51
  evolution, 9
  head anatomy, 16–17
  natural history, 50–53, 77, 83
  population estimate, 176
  synonyms, 49
  taxonomic notes, 54
Cyclura carinata
  Booby Cay population, 10
  phylogenetic tree, 5
Cyclura carinata carinata, *see* Turks and Caicos iguana
Cyclura collei, *see* Jamaican iguana
Cyclura cornuta, phylogenetic tree, 5
Cyclura cornuta cornuta, *see* Rhinoceros iguana
Cyclura cornuta onchiopsis, *see* Navassa Island iguana
Cyclura cornuta stejnegeri, *see* Mona Island iguana
Cyclura cychlura cychlura, *see* Andros Island iguana
Cyclura cychlura figginsi, *see* Exuma Island iguana
Cyclura cychlura inortata, *see* Allen Cays iguana
Cyclura cychlura
  evolution, 9
  phylogenetic tree, 5
Cyclura lewisi, *see* Grand Cayman Blue iguana
Cyclura nubila
  evolution, 6
  phylogenetic tree, 5
Cyclura nubila caymanensis, *see* Sister Isles iguana
Cyclura nubila nubila, *see* Cuban iguana
Cyclura pinguis, *see* Anegada Island iguana
Cyclura ricordii, *see* Ricord's iguana
Cyclura rileyi
  evolution, 9
  phylogenetic tree, 5
Cyclura rileyi cristata, *see* White Cay iguana
Cyclura rileyi nuchalis, *see* Acklin's iguana
Cyclura rileyi rileyi, *see* San Salvador iguana
Cystic calculi, management, 159–160

Diet
  determination, 129–130, 132–136
  forage and food types, 80–82, 131
  husbandry, *see* Feeding
  nutrients
    analysis, 132–133
    requirements, 132, 134
  seasonal changes, 139–140
Digestion, 80–82
Dog
  control, 188
  iguana predation, 11, 84–86, 171, 180
Dystocia, management, 155–157

Ecotourism, iguana impact, 184
Education programs, 196–199
Eggs
  clutch size, 91
  handling and incubation, 115–117
  oviposition, 91, 115
Euthanasia, 165
Evolution
  biogeography, 5–9
  island distribution of iguanas, 3–5
  phylogenetic tree, 5
Exuma Island iguana
  conservation status, 37
  description, 34
  natural history, 35–37
  population estimate, 176
  synonyms, 34
  taxonomic notes, 37

Feces, feeding, 80, 82
Feeding
  availability, 137–138
  browse plants, 106–108
  daily diet composition
    adults, 107–108
    juveniles, 109
  food dishes, 104–105
  fighting, 105
  hatchlings, 105
  health concerns
    gout, 141
    metabolic bone disease, 140–141
  nutrient analysis, 136–138
  preparation of greens, 106, 138
  reproduction effects in females, 154

Reptile Carnovore/Omnivore Gel, 107–108
serum vitamin data, 143
stomach tubing, *see* Gavage feeding
vitamin D
oral supplementation, 142
status assessment, 142
synthesis, 141
Fighting behavior, 87–89
Fish, iguana predation, 84
Follicular stasis, management, 154–157
Food competition, domestic animals and iguanas, 86, 171
Fossils, rock iguana species, 72–73

Gavage feeding
formula dosing, 165
indications, 168
technique, 165
tools, 163
Gout, dietary concerns, 141
Grand Cayman Blue iguana
breeding, 112
conservation status, 45
description, 42–43
diet seasonal changes, 139–140
distribution, 5
evolution, 10–11
hatchlings, 119–121
headstarting, 189–192
natural history, 43–45
phylogenetic tree, 5
population estimate, 176
synonyms, 42
taxonomic notes, 45
taxonomy, 9–10
Growth rate, 91–93

Habitat
loss and degradation, 178–179
protection, 185–187
requirements, 77
Handling, captive iguanas, 108, 110–112, 154
Hatchling
care in captivity, 117–123
dispersal, 91
dominance hierarchy, 121
feeding, 105
handling, 110–111
predation, 84–87, 91
size, 91
Head
anatomy, 16–17
bobbing, 87–89
Headstarting
health assessment of iguanas, 166–167
overview, 189–192
Health assessment, free-range and headstarting iguanas, 166–167
Hemipene prolapse, management, 157
Hispaniola, rock iguana evolution, 7
Home range, 78–79
Housing
burrowing, 99–100
examples, 99–102
hatchlings and juveniles, 121–122
materials, 100
plants, 99
size, 98–99
temperature, 103–104
ultraviolet light, 103–104, 121–122
Humans
activity interactions with iguanas, 183–185
iguana predation, 11, 84, 182–183
Hunting, 182
Husbandry
breeding and nesting, 112–117
feeding, 104–108
handling, 109–112
hatchling care, 117–123
housing, 98–104
medical survey, 148
population management, 97–98
quarantine, 98
record keeping, 123–125

Iguana Specialist Group, 196, 198–199
IIF, *see* International Iguana Foundation
Infection, types and management, 153–154, 170
Insects, feeding, 82
International Iguana Foundation (IIF), 194
International Reptile Conservation Fund (IRCF), 194, 198
Intestine
bacterial flora, 168, 170
colon complexity, 80
impaction, 158–159
IRCF, *see* International Reptile Conservation Fund

Jacobson's organ, 87
Jamaican iguana
conservation status, 24
description, 21–22
evolution, 7

Jamaican iguana (*Continued*)
habitat protection, 185–187
headstarting, 189
natural history, 22–24
phylogenetic tree, 5
population estimate, 176
synonyms, 21
taxonomic notes, 24

Kidneys, *see* Renal disease

Leukocytosis, management, 161
Life span, 92–93

Metabolic rate, rock iguana, 80
Mites, management, 150–151, 170
Mona Island iguana
conservation status, 30
description, 28
headstarting, 189–192
natural history, 29–30
population estimate, 176
synonyms, 28
taxonomic notes, 30
Mongoose, iguana predation, 11, 86, 180

Navassa Island iguana
conservation notes, 72
description, 71–72
natural history, 72
synonyms, 71
taxonomic notes, 72
Nest box, 114
Nesting behavior, 90–91, 113–114
Nutrition, *see* Diet; Feeding

Oophoritis, management, 154
Outreach programs, 196–199
Oviductal prolapse, management, 157
Oviposition, 91, 115

Parasites, *see also specific parasites*
types and management, 149–153, 170–171
Passive integrated transponder (PIT), implantation, 98, 120–122
Pathology, rock iguana, 171–172
Pig
control, 188
iguana predation, 85, 181–182
Pinworm, management, 151–152, 170–171
PIT, *see* Passive integrated transponder
Poaching, 182–183
Population and Habitat Viability Workshop, 199
Predation
defenses, 84–87
predators of iguanas, 11, 84–86, 180–182
Protozoa infection, management, 152

Quarantine, 98

Raccoon, iguana predation, 85
Radiotelemetry, 196
Rat, iguana predation, 85, 181–182
Record keeping, husbandry, 123–125
Renal disease, management, 157–158
Reproduction
breeding husbandry, 112–117
clutch size, 91
copulation, 112–113
courtship behavior, 89–90
diseases, 153–157
egg handling and incubation, 115–117
nesting behavior, 90–91
oviposition, 91, 115
season, 112
senescence, 93
Reptile Carnovore/Omnivore Gel, 107–108
Restraint, captive iguanas, 108, 110–112
Rhinoceros iguana
breeding, 112
conservation status, 27
description, 25
distribution, 7
natural history, 25–27
population estimate, 176
synonyms, 24
taxonomic notes, 27
Ricord's iguana
conservation status, 62
description, 59–60
distribution, 7
habitat protection, 185–186
natural history, 60–61
phylogenetic tree, 5
population estimate, 176
synonyms, 59
taxonomic notes, 61
Roads, iguana fatalities, 185

Salpingitis, management, 154
Salt gland, 80
San Salvador iguana
conservation status, 71
description, 68–69
natural history, 69–70
population estimate, 176

synonyms, 68
taxonomic notes, 70
Sauroplasma infection, 171
Sex ratio, 120
Sexing, hatchlings, 120
Sister Isles iguana
conservation status, 49
description, 46–47
natural history, 47–48
population estimate, 176
synonyms, 46
taxonomic notes, 49
Small Population Animal Record Keeping System (SPARKS), 97
Snakes, iguana predation, 84, 86, 91
Snout-to-vent length, *see* Body measurements
Social behavior, 87–89
SPARKS, *see* Small Population Animal Record Keeping System
Species, definition, 9, 86
Species Survival Plan (SSP)
AZA Rock Iguana Species Survival Plan, 194–195
goals, 195
rock iguana species targets, 195
SSP, *see* Species Survival Plan
Stomach tubing, *see* Gavage feeding

Tapeworm, management, 151
Taxonomy, *Cyclura*, 9–10
Teeth, 81
Ticks, management, 149–150, 170
Traffic, iguana fatalities, 185
Translocation, programs for iguanas, 192–193
Trauma, types and management, 148–149
Turks and Caicos iguana
cannibalism, 83
conservation status, 21
description, 17
evolution, 8
natural history, 17–20, 80, 82–83
population estimate, 176
synonyms, 17
taxonomic notes, 20

Ultraviolet light
husbandry, 103–104, 121–122
vitamin D synthesis, 134, 141–142

Vitamin A, status, 142–143
Vitamin D
deficiency and metabolic bone disease, 140–141, 158
oral supplementation, 142
status assessment, 142, 171
synthesis, 136, 141–142

White Cay iguana
conservation status, 62
description, 62–63
natural history, 63–64
population estimate, 176
synonyms, 62
taxonomic notes, 64

Yolk coelomitis, management, 157

Printed and bound by CPI Group (UK) Ltd, Croydon, CR0 4YY
03/10/2024
01040332-0003